Marine microbiology

Marine microbiology

B. AUSTIN

Department of Brewing and Biological Sciences
Heriot-Watt University

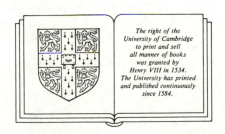

The right of the
University of Cambridge
to print and sell
all manner of books
was granted by
Henry VIII in 1534.
The University has printed
and published continuously
since 1584.

CAMBRIDGE UNIVERSITY PRESS

Cambridge

New York   New Rochelle   Melbourne   Sydney

Published by the Press Syndicate of the University of Cambridge
The Pitt Building, Trumpington Street, Cambridge CB2 1RP
32 East 57th Street, New York, NY 10022, USA
10 Stamford Road, Oakleigh, Melbourne 3166, Australia

© Cambridge University Press 1988

First published 1988

Printed in Great Britain at the University Press, Cambridge

British Library cataloguing in publication data

Austin, B.
  Marine microbiology.
  1. Marine microbiology
  I. Title
  576′.192   QR106

Library of Congress cataloguing in publication data

Austin, B. (Brian). 1951–
  Marine microbiology/B. Austin
  p.   cm.
  Includes bibliographies and index.
  ISBN 0 521 32252 9.   ISBN 0 521 31130 6 (paperback)
  1. Marine microbiology.   I. Title.
QR106.A97 1988
576′.192–dc19   87-25373   CIP

ISBN 0 521 32252 9 hard covers
ISBN 0 521 31130 6 paperback

MTL

To Dawn Amy Austin for her patient supportive
role in the preparation of this book

Contents

Preface   xi

1   **Introduction**   1
1.1   The marine environment   1
1.1.1   Seawater   4
1.2   Marine sediments   8
1.3   Habitats for marine micro-organisms   9

2   **Microbiological methods**   12
2.1   Sampling   12
2.1.1   Water samples   13
2.1.2   Sediment samples   13
2.1.3   Aquatic contents   13
2.2   Microscopic examination of the samples   15
2.3   Determination of biomass   19
2.4   Culturing techniques   23
2.4.1   Prokaryotes   23
2.4.2   Eukaryotes   29

3   **Quantification of marine microbial populations**   31
3.1   Prokaryotes   32
3.1.1   Numbers of bacteria in the marine environment   32
3.1.2   Quantitative data of specific groups of
          prokaryotes in the marine environment   36
3.1.3   Influence of methods on the estimation of bacterial popula-
          tions   40
3.1.4   Biomass   42
3.2   Eukaryotes   42

4   **Taxonomy of marine micro-organisms**   45
4.1   Prokaryotes   47
4.1.1   Phototrophs containing bacteriochlorophyll   47
4.1.2   Cyanobacteria   51
4.1.3   *Prochloron*   52

Contents

4.1.4	Gliding bacteria	53
4.1.5	Budding and appendaged bacteria	54
4.1.6	Aerobic Gram-negative rods and cocci	55
4.1.7	Facultatively anaerobic Gram-negative rods	58
4.1.8	Gram-negative anaerobic rods and cocci	59
4.1.9	Gram-negative chemolithotrophs (ammonia- or nitrite-oxidising bacteria)	60
4.1.10	Gram-negative chemolithotrophs (sulphur bacteria)	61
4.1.11	The methane bacteria	62
4.1.12	Aerobic Gram-positive cocci	63
4.1.13	Gram-positive endospore-forming rods	64
4.1.14	Actinomycetes and related bacteria	65
4.1.15	The spirochaetes	66
4.2	Eukaryotes	66
4.2.1	Micro-algae	66
4.2.1.1	*Diatoms*	66
4.2.1.2	*Other micro-algae*	68
4.2.2	Filamentous fungi	68
4.2.3	Yeasts	70
4.2.4	Protozoa	70
5	**Ecology**	79
5.1	Survival of indigenous organisms in the marine environment	79
5.2	Fate of non-indigenous organisms in the marine environment	82
5.3	Predator-prey relationships (food webs)	83
5.4	Degradation of complex molecules	87
5.5	Colonisation of surfaces, and the role of chemotaxis and attachment	88
5.5.1	Chemotaxis	88
5.5.2	Attachment	89
5.6	Biogeochemical processes (nutrient cycling)	90
5.6.1	The nitrogen cycle	90
5.6.2	The sulphur cycle	93
5.6.3	Carbon cycling	96
5.6.4	Other cycles	99
5.7	Primary productivity	99
6	**Microbiology of macro-organisms**	103
6.1	Microbiology of plants	104
6.2	Microbiology of healthy vertebrates	105
6.3	Microbiology of healthy invertebrates	111
6.4	Diseases of vertebrates	115

Contents

6.4.1	Bacterial pathogens	118
6.4.2	Fungi	120
6.4.3	Protozoa	121
6.4.4	Viruses	122
6.5	Diseases of invertebrates	123
6.5.1	'Vibriosis'	123
6.5.2	Shell-disease	127
6.5.3	Gaffkemia	127
6.5.4	Epibiotic associations	129
6.5.5	Fungal diseases	129
6.5.6	Virus diseases	130
6.5.7	Rickettsial diseases	131
7	**Microbiology of the deep sea**	133
7.1	The deep-sea environment	134
7.1.1	Hydrothermal vents	135
7.2	Experimental approach to the study of deep-sea microbiology	135
7.3	The microflora of the deep sea	138
7.3.1	Quantitative aspects	138
7.3.2	Quantitative aspects – hydrothermal vents	140
7.4	Taxonomy of deep-sea microbial populations	141
7.5	Activity and role of deep-sea micro-organisms	146
7.5.1	Chemosynthetic activity at deep-sea hydrothermal vents	148
8	**Benefits and malefits of marine micro-organisms**	149
8.1	Benefits	149
8.1.1	Biodegradation of pollutants	149
8.1.2	Role of micro-organisms in the settling of invertebrate larvae	152
8.1.3	Microbial involvement with the formation of manganese nodules	152
8.1.4	Fermented food products	153
8.2	Malefits	153
8.2.1	Biodeterioration/biofouling of objects in the sea	153
8.2.2	Mobilisation of heavy metals	154
8.2.3	A reservoir for human pathogens	154
8.2.4	Spoilage and food-poisoning micro-organisms in fish	156
9	**Biotechnology**	157
9.1	Pharmaceutical compounds	157
9.1.1	Antibiotics	157
9.1.2	Antiviral compounds	170
9.1.3	Antitumour compounds	171

Contents

9.2   Enzymes   172
9.3   Surfactants   173
9.4   Other potentially useful microbial products   173

10   Future developments   **175**

       References   177
       Index   218

Preface

There has been a steady momentum of research activity in marine microbiology since the pioneering efforts of Dr C.E. Zobell. Indeed, his treatise, entitled *Marine Microbiology* (published in 1946) has been a source of inspiration for students and established scientists alike. My initial exploration of the text resulted from an interest in chromogenic bacteria, albeit of terrestrial leaf-surface (phylloplane) origin. A computer-based literature search revealed the presence of articles detailing chromogens in the marine environment. Thus, a few fruitful days were spent, closeted in the library, reading Zobell's book. In 1972, I regarded the marine environment to be far removed from leaf surfaces of terrestrial plants. However, my interests also encompassed numerical taxonomy; a topic which was studied further during postdoctoral work in the laboratory of Dr R.R. Colwell. The jump to aquatic microbiology followed during this period at the University of Maryland. Then, specialisation followed in the fascinating topic of fish diseases at the Fish Diseases Laboratory in Weymouth.

Several years later after joining the academic staff of Heriot-Watt University, I met Dr Fay Bendall of Cambridge University Press in the auspicious building of the Royal Society in London. After outlining my teaching programme in marine microbiology and emphasising the dearth of modern texts on the subject, an invitation was received to prepare a manuscript. Thus, the present book was conceived. However, after commencing the task, a third edition of Professor G. Rheinheimer's excellent text entitled *Aquatic Microbiology* was published. My plans were hastily revised.

The primary aim has been to prepare a concise text detailing the current understanding in marine microbiology. However, information about estuarine ecosystems has been largely ignored, insofar as the topic deserves a separate book in its own right. Moreover, the book is directed at newcomers in the field, notably undergraduates and young research workers. Emphasis has been placed on new and exciting developments such as those related to biotechnology, fish and shellfish pathology, and the concept of dormancy. Some of the more well established and documented facets of marine microbiology have been condensed in a space-saving gesture.

Grateful thanks are extended to Drs N. MacDonald and R.A. Herbert for supplying figures. Much of the photography was capably carried out by Mr James Buchanan. Finally, efficient secretarial assistance was provided by Mrs M.A.M. Dunn and by my wife, Dr D.A. Austin, the latter of whom also willingly helped with the editing of the typescript.

B. Austin

Edinburgh 1987

1   Introduction

There is a growing awareness of the role of marine micro-organisms in biogeochemical processes, for biotechnology, in pollution, and in disease. For example, there has been increasing interest in the role of methanogens in sediments. This work has produced some excellent publications on ecology and taxonomy – topics which will be considered in later chapters. However before detailed consideration of marine micro-organisms is made, it would be relevant to discuss the nature of the environment.

Planet Earth is dominated by the seas, which cover approximately 71% of its surface. The role of the seas to mankind includes recreation as well as commerce. Coastal areas in temperate and tropical regions serve for recreation, e.g. summer holidays by the seaside. Offshore, thriving industries contribute fisheries products, petro-chemicals and minerals to the socio-economic well-being of the community. The exploration for oil in the North Sea and the Gulf of Mexico is a growth industry contributing substantially to employment and the wealth of many nations. The over-exploitation of fish stocks is of concern at an international level, with the resultant formulation of treaties aimed at conservation of selected species. The emotive issues of pollution further strengthen public awareness of the problems of the seas. The influxes of organic materials, such as untreated sewage and radioactive waste from terrestrial sites into the coastal areas stimulate heated debate with the overriding concern for public health. Deliberate or accidental contamination of large areas of the sea with oil slicks, usually from large ships, has been the subject of debate for several decades.

1.1   The marine environment

Seas comprise the largest continuous expanses of water on the planet. In total, the seas cover 3.61×10^8 km^2 with an average depth of 3800 m, a total estimated weight of water of 1.41×10^{18} tonnes and a volume of 1.4×10^{21} l. Expressed another way, the seas contain 97% of all the known water on Earth, most of which is in the Southern Hemisphere.

On a global basis, there are three interconnected oceans, i.e. the Atlantic, Indian and Pacific, and many seas, namely the Arctic, Bering, Black, Caspian, Japan, Mediterranean, Okhotsk, Ross and Weddell (Fig. 1.1). In terms of size, the Pacific is the largest ocean, being approximately equal to the combined area of the Atlantic and Indian Oceans. Therefore, it is hardly surprising that the Pacific contains more than half of the world's water. Moreover, the Pacific Ocean has the greatest average depths, i.e. 3940 m, and the deepest trenches of approximately 11 000 m (Table 1.1). This compares to average depths in the Atlantic and Indian Oceans of 3310 m and 3840 m, respectively.

Starting at the coastal regions and working outwards, the seas gradually increase in depth before abruptly sloping downwards. The zone which joins the coast is referred to as the *continental shelf*, and reaches a maximum depth of approximately 200 m. The continental shelf extends for a width of approximately 65 km before the downward slope (known as the *continental slope*) leads into the deep water of the ocean basins. These basins, with depths of water of 4000–6000 m, occupy almost one-third of the Earth's surface. The sea floor is often portrayed as being relatively flat, however, this is far from correct. Long, narrow sheer-sided trenches and high mountains or volcanoes lie submerged beneath the water. In fact, there are nearly 10 000 volcanoes on the sea floor, some of which break above the surface of the water to give rise to volcanic islands, e.g. the Hawaiian Islands. Geological faults have also been detected at the bottom of the oceans. In addition, there are wide plains, i.e. the *Abyssal Plains*.

The seas are not static in size or shape, as superficial glances at their ever-changing shorelines should reveal. Thus, in some areas, the seas advance and erode away the shore by means of wave action, whereas in other locations they retreat. Extensive changes have also resulted from glaciation. Violent storms and giant waves, e.g. tsunamis, may also occur.

Table 1.1. *Maximum depth of the deep-sea trenches*

Ocean	Trench	Depth (m)
Atlantic	Puerto-Rican	9 200
	South Sandwich	8 400
Indian	Javan	7 460
Pacific	Aleutian	8 100
	Japan	9 800
	Kermadec–Tongan	10 800
	Kuril–Kamchatkan	10 500
	Marianas	11 000
	Peru–Chile	8 050
	Philippine	10 000

Data from Grant Gross (1976).

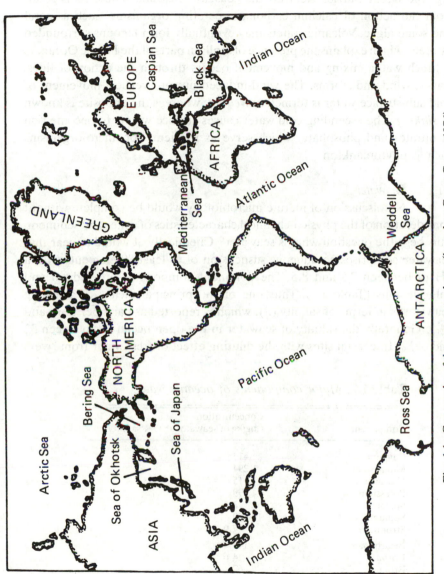

Fig. 1.1 Geography of the oceans and seas (based on Grant Gross, 1976).

The latter often result through submarine volcanic activity. In the Indian and Pacific Oceans, particularly near the warmer waters of the equator, the sites of ancient volcanoes are indicated by the presence of coral reefs, e.g. the Great Barrier Reef off the coast of Australia. These reefs result from the action of calcium carbonate-secreting organisms, such as coral and some algae. Volcanic craters may eventually form lagoons surrounded by reefs, which explains the presence of atolls in parts of the Pacific Ocean.

Much water mixing and movement occurs through the action of tides, waves, wind and storms. The wind-induced upward vertical movement of cold sub-surface water is termed *upwelling*, whereas the opposite is known as *sinking*. The ascending cold water causes surface waters to become rich in nitrates and phosphate, which serve as nutrients for micro-organisms such as phytoplankton.

1.1.1   *Seawater*

No discussion of marine microbiology would be complete without consideration of the physico-chemical characteristics of water and sediment. This begs the question what is seawater? Chemically, it would appear that seawater comprises an aqueous suspension of at least 80 elements with a pH of between 7.5 and 8.5. The dominant components are sodium and chloride ions (Table 1.2). Thus one of the principal means of assessing seawater is in terms of salinity (S), which is reported as parts per thousand (‰). Generally the salinity of seawater in the open ocean lies between 32 and 38‰. In coastal sites with the diluting effect of freshwater from rivers

Table 1.2. *Major components of oceanic water*

Component	Concentration (mg/kg of seawater at 35‰ salinity)
Calcium	412
Magnesium	1 294
Nitrogen	15
Potassium	399
Silicon	2.9
Sodium	10 760
Strontium	7.9
Bicarbonate	145
Boron	4.6
Bromide	67
Chloride	19 350
Fluoride	1.3
Sulphate	2 712

Based on Pytkowicz and Kester (1971).

and land run-off, the salinity is often lower, and lies in the range of 10–32‰. Of course, changes in salinity will result from evaporation, in which case freshwater is removed as water vapour with a concomitant rise in salinity. Conversely, precipitation in the form of rain or snow exerts a diluting effect. High salinities have been recorded in the Red Sea where values of 44‰ are not uncommon. At the polar regions, the removal of freshwater through the formation of ice also produces high salinities.

Apart from sodium and chloride ions, seawater contains large quantities of calcium, magnesium, nitrogen, potassium, silicon, strontium, bicarbonate, boron, bromide, fluoride and sulphate (Table 1.2). Apparently, most of the calcium, as calcium carbonate, is derived from organisms. The total input of calcium is in the region of 7×10^{14} g/year, which has the potential to form 1.8×10^{15} g of calcium carbonate. Smaller quantities of approximately 60 other elements have been found in oceanic water (Table 1.3).

Table 1.3. *Minor components of oceanic water*

Component	Concentration (μg/kg of seawater at 35‰ salinity)
Aluminium	2
Argon	4
Arsenic	3
Barium	20
Iron	2
Iodide	60
Lithium	180
Molybdenum	10
Phosphorus	60
Rubidium	120
Titanium	1
Uranium	3
Vanadium	2
Zinc	3
Cadmium, chromium, caesium, copper, krypton, manganese, neon, nickel, antimony	2.4[a]
Bismuth, cobalt, gallium, germanium, mercury, niobium, lead, selenium, tin, thallium, xenon, zirconium	0.4[a]
Silver, gold, beryllium, cerium, helium, hafnium, lanthanum, tantalum, thorium, tungsten, yttrium, radium	0.05[a]
Dysprosium, erbium, europium, gadolinium, holmium, lutetium, praseodymium, samarium, terbium, thulium, ytterbium	0.005[a]
Indium, protactinium, scandium	0.0007[a]

[a]Concentrations are sum totals for all elements.
Based on Holland (1978).

Some of these elements are transferred from rivers, for example, adsorbed to clay particles. Others result from the chemical reaction of water with oceanic basalts. It should be emphasised that from an historical perspective, there is negligible information concerning possible changes in the composition of seawater with time. However, it may be assumed that with the advent of the Industrial Revolution, the deposition of organic and inorganic pollutants in the marine environment, has increased. Such additions may exert dramatic impacts on marine micro-organisms.

With such large surface areas, the seas are greatly influenced by solar energy. The visible spectrum of light is able to penetrate to a depth of no more than 200 m, although infra-red radiation is unable to penetrate more than 1 m. Within this 200 m zone, known as the *photic zone*, photosynthesis occurs. All primary productivity takes place here and often the literature will refer to this as the *productive zone*. Of course, one of the products of photosynthesis is oxygen, which contributes to the amount of dissolved oxygen in seawater. The depth of the photic zone reflects the tubidity of the water. Thus, the depth of the photic zone is inversely related to the amount of particulate matter in the water. In some coastal sites, this zone may be less than 1 m in depth. Waters, in which light does not penetrate, are referred to as *aphotic*. Here, oxygen-generating photosynthesis is unable to proceed.

Essentially, the adsorption of solar energy produces three layers within the water column; referred to as the *surface*, *pycnocline* and *deep zones*. At the surface there may be marked seasonal variations in the heating, cooling and evaporation of the water. These effects are particularly noticeable in coastal locations of temperate regions. In the open ocean, incoming solar energy causes the evaporation of water, with the concomitant formation of rain. Consequently, this has a dramatic influence on coastal sites, insofar as rain, being freshwater, may have a diluting effect upon the chemical composition of seawater. Alternatively, rain inevitably leads to the run-off of organic and inorganic material from the land. At polar and sub-polar regions, the freezing of surface water routinely occurs, with a dramatic influence on both salinity and density. As the resultant ice is largely freshwater, the salinity of the underlying liquid water increases with a concomitant rise in density. Generally, surface waters are less dense than those beneath. However, much vertical mixing of the water occurs in the surface zone.

Beneath the surface waters, a zone occurs in which the water density changes with depth. This is the slope known as the pycnocline, which functions as a fairly effective barrier to vertical water movement. In the deep zone, which contains approximately 80% of the seawater, large masses of water occur and move. These waters are recognisable by salinity and temperature.

Although the surfaces of the open ocean are subject to seasonal changes in sunlight, there is a marked uniformity of temperature in the surface waters of many regions. In fact, the variation is often not more than 0.2–0.3 °C. However, less heat reaches the planet at the poles, therefore, there are significant differences in the surface water temperatures of polar, temperate and tropical regions. In the last, temperatures of 25–32 °C are normal with a variation of less than 2 °C, whereas the coldest water occurs at the poles. Here, the average temperature is only 3.5 °C, but it may decline to −1.7 °C in the ice containing areas. In such locations, the variation in temperatures usually changes by 5–9 °C. Thus at polar sites, psychrophiles (i.e. organisms which grow at up to 10–15 °C) predominate, whereas mesophiles (i.e. organisms which grow in a temperature range of 15–40 °C) occur in large numbers in the tropics.

Solar radiation also causes a layered structure in the temperature of the water. The *thermocline* is the region in which the water temperature changes abruptly with depth. Below the thermocline, there is minimal change in the temperature despite an increasing depth. The temperature changes of the thermocline cause subtle changes in water pressure. Thus in open oceans, the thermocline inevitably coincides with the pycnocline.

The cold water of the deep sea originates primarily at the poles, most notably around Antarctica. There, the surface waters are sufficiently cold to permit the formation of sea ice, with the resultant exclusion of salts during the freezing process. These salts become mixed with the underlying cold water, thereby increasing salinity. This action leads to the formation of cold dense waters which sink along the Antarctic continental shelf and Antarctic continental slope to reach the bottom of the deep sea. Here, the water interacts with a current which circles Antarctica and mixes with the adjacent water. The resultant water mass, known as the Antarctic Bottom Water with a temperature of −0.4 °C and salinity of 34.66‰ flows northward into all three ocean basins. Sub-surface water masses similarly form in the North Atlantic and North Pacific, but whereas such dense waters are important locally they do not reach all three oceans. The North Atlantic Deep Water, with a salinity of 34.9–35.0‰ and temperature of 3–4 °C, is one of these dense masses of water which sinks to the bottom of the ocean, flows in a southerly direction, and eventually passes over the denser Antarctic Bottom Water. Antarctic Intermediate Water (temperature of 3–5 °C; salinity of 34.1–34.6‰) is a mass of water of intermediate density which forms around Antarctica. It is denser than surface water, below which it flows in a northward direction. These water masses assume relevance in subsequent discussion of the microbiology of the deep sea.

1.2   Marine sediments

A large proportion of the sea bottom is covered with sediment, which contains rock fragments and animal debris, namely pieces of skeleton and shell. The depth of this sediment varies considerably with geographical area being approximately 600 m thick in the Pacific Ocean (this is regarded as the average thickness of deep sea sediment), increasing to >9 km in thickness within the Puerto Rican Trench. In comparison, there is a 500–1000 m thick layer of sediment in the Atlantic Ocean. However, within the colder Arctic Sea, the sediment may be >4 km in thickness, of which approximately 50% is water.

Sediments may be described as 'sand', if individual particles exceed 62 μm in diameter, or 'mud' (if particles are <62 μm in diameter). Sand particles tend to be deposited on the Continental Shelf, and commonly occur along shorelines as beaches. In deep-sea sediments, sand often comprises <10% of the components. Here the dominant constituent is mud.

Apart from describing marine sediments as muds or sands, the particles may be characterised by origin, namely:

(a) *biogenous*, this is a type of sediment particle which is derived from organisms, e.g. shell debris. Essentially, biogenous particles may be rich in calcium carbonate (calcareous), silicate (siliceous) or phosphate (phosphatic) depending upon whether they are derived from the shells of foraminifers or diatoms, or from the bones/scales of marine vertebrates. To be described as biogenous, a sediment must contain more than 30% of biogenous particles. Such sediments are common, with calcareous muds covering half of the bottom of the deep sea. These sediments accumulate at a rate of 1–4 cm/1000 years.

(b) *hydrogenous* particles originate from water derived from inorganic chemical reactions; a topical example is the occurrence of manganese nodules which cover large areas of the Pacific floor. Formation of such nodules is quite slow, being measured in terms of <0.01–1.0 mm/1000 years.

(c) *lithogenous* particles, which result from the erosion of rocks, with the particles being transported to the sea via rivers. The particles are deposited on the sea floor when the movement of water is insufficient to support them. On the continental slope, the particles are sorted by size according to the action of waves and currents. The largest particles are deposited (or retained) whereas the smallest are swept away. Generally, the largest particles accumulate near shorelines whereas the smallest grains move seaward and are eventually deposited in deep water. In such locations lithogenous sediments accumulate at a rate of approximately 1 mm in thickness/1000–10 000 years. However, at the mouths of large rivers, close to shore, and as a result of the action of glaciers, the sediments accumulate much faster, with rates of several metres in thickness per thousand years not being uncommon. Within the sediments, there may be a variety of chemical reactions taking place, which give rise to certain recognisable characteristics. For example, the interaction of dissolved oxygen with iron leads to the formation of iron oxides, which are readily apparent in red clays. There may be substantial mixing and movement of sediments, such as may occur as a result of earthquakes. Thus, underwater sediments comprise a complex dynamic environment for micro-organisms.

1.3   Habitats for marine micro-organisms

There are eight basic kinds of habitats for marine micro-organisms (Fig. 1.2). At the surface of the sea, there is the *neuston* (also known as *pleuston*). This is a term which was initially described by Naumann (1917) for the polysaccharide–protein rich micro-habitat at the air–water interface. *Plankton* comprise poorly-motile organisms, many of which are photosynthetic, suspended largely in the photic zone of the water column. The plankton may be described in terms of composition, i.e. plant (phytoplankton), animal (zooplankton) or bacteria (bacterioplankton), or by size, namely femtoplankton (0.02–0.2 μm), picoplankton (0.2–2.0 μm), nanoplankton (2.0–20 μm), microplankton (20–200 μm), mesoplankton (0.2–200 mm), macroplankton (20–200 mm) and megaplankton (200–2000 mm). On this basis, the bacterioplankton would be in the size range of picoplankton.

The *nekton* comprise the large swimming animals, e.g. fish, which consume the plankton. These animals produce organic debris, referred to as *seston*, which contributes significantly to nutrition of the micro-organisms. The simple organic compounds are readily broken down in the water column, whereas larger particles or those more resistant to biodegradation sink to the sea floor. Here, the material may serve as a nutrient source to bottom dwelling organisms.

Epibiotic habitats are inanimate (e.g. biofouling communities) or animate surfaces on which attached communities occur. Conversely *endobiotic* habitats appertain to the environment within the tissues of other larger organisms. Here, the relationship with the host may be beneficial (*mutualism*), detrimental to the host insofar as the invading organism derives food but does not necessarily cause disease (*parasitism*), or harmful insofar as a disease situation results (*pathogenesis*).

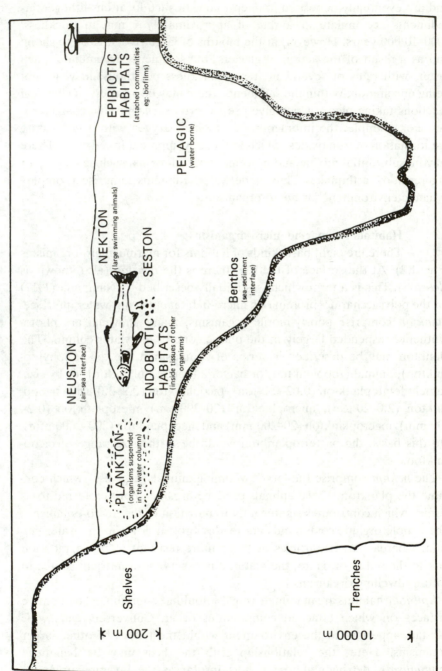

Fig. 1.2 Schematic representation of the habitats for marine micro-organisms (based on Sieburth, 1979).

Water-borne habitats are referred to as *pelagic*, although the term may be more tightly defined according to the precise depth in the water column. Thus starting at the surface and descending to the bottom waters, there are *epipelagic* (extending from the surface to a depth of approximately 100 m; i.e. the photic zone), *mesopelagic, bathypelagic* and *abyssopelagic* habitats. Finally, life on the seafloor, i.e. at the sea–sediment interface comprises the *benthos* (also known as the *benthic* habitat). Within these habitats, all the complex interactions of marine micro-organisms ensue.

2   Microbiological methods

Information about the size and nature of marine microbial populations reflects the methodology used, encompassing sampling procedures, the techniques used to assess biomass and the taxonomic composition of the microflora.

2.1   Sampling

It may be correctly assumed that the difficulty with which samples are obtained increases proportionally with the distance from shore. At the shore, samples of water and sediment may be readily collected in sterile bottles, whereas plankton and fish may be captured in hand-held plankton nets and on fishing lines, respectively. In the shallow coastal waters, up to a few hundred metres from shore, there will be a need for a small boat, such as a rowing boat, rubber dinghy or small motor boat. Again, the collection equipment may be quite simple although bottles may need to be suspended from lines, and sediment samples may have to be obtained via grabs. Further from shore, large research vessels assume importance, and the sampling equipment may be quite complex. For deep-sea samples, submersible vessels are often needed. At this point, the financial commitment becomes very large, with the hire of surface vessels costing many thousands of pounds (sterling) per day. Often there will be a need to supply all equipment, even gas burners, and the neglect (or loss) of even a small item may negate the value of a research cruise. Much attention to detail is needed, particularly before the onset of the cruise. Such details include verification of the nature of the electrical supply onboard ship, insofar as American 110 volt equipment often does not work well off the British 220 volt system, unless suitable adaptors are available. Moreover, less controllable factors may affect the sampling programme, including the weather, loss of equipment overboard, and seasickness among the personnel. Samples to be collected will inevitably include a variety of water, sediment and aquatic contents, e.g. fish and plankton.

2.1.1 *Water samples*

To collect neuston, an effective method used by the author is to use sterilised sheets of tissue paper, which are positioned on the surface of the water. This operation usually requires the use of a low-sided boat and a comparatively still water surface. After a pre-determined interval, of for example 5–10 minutes, the tissue paper is carefully removed by means of forceps, drained of excess water, and placed in a suitably sized container. The organisms may be subsequently removed by shaking in diluent. Other samplers have been devised, including a rotating drum which skims off the top layer of water (Fig. 2.1) and a hand-held net (Fig. 2.2).

Non-sterile water samples may be collected in Nansen bottles (Fig. 2.3). These cylindrical vessels, with a capacity of several litres, have rubber seals at the open ends. The bottles are attached to lines, lowered to pre-determined depths, and the triggering mechanism released via metal weights (known as messengers) dropped down the line. Thus, the rubber discs are pulled inward, forming a tight waterproof seal. Hence the samples may be raised, and water removed via taps.

For sterile water, the Niskin sampler is ideal. This involves a pre-sterilised plastic bag, which fits onto a winged sampler (Fig. 2.4). The wings are locked together, fitted with the plastic bag, and positioned on a line. At the pre-determined depth, a messenger operates a guillotine which opens a hose into the bag and simultaneously releases the wings of the sampler. These wings spring apart, drawing water into the bag. The bag may now be retrieved for subsequent examination.

Deep-sea samples may be retrieved in pressurised containers, which permit the recovery of water at *in situ* pressures. Alternatively, the use of submersible vessels, e.g. the Alvin, may be used (Fig. 2.5).

2.1.2 *Sediment samples*

Sediment may be collected via corers. These descend rapidly on a line, and bury in the sediment. Alternatively, the use of grabs permits the recovery of portions of the upper layers of sediment (Fig. 2.6(*a*)(*b*)). However, there may be justifiable concern about microbial contamination of the samples.

2.1.3 *Aquatic contents*

Epibiotic and benthic communities may be collected by means of SCUBA equipment. The former may also be obtained by immersing inert structures, e.g. glass slides, into the sea for pre-defined periods. The latter are also amenable to collection by dredging (Fig. 2.7). Nets (Figs. 2.8, 2.9)

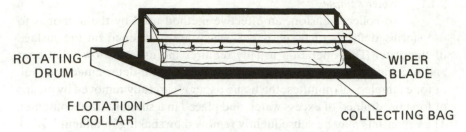

Fig. 2.1 A rotating drum used to sample neuston (after Sieburth, 1979)
Photograph courtesy of Dr R.A. Herbert (reproduced with permission of
John Wiley & Sons)

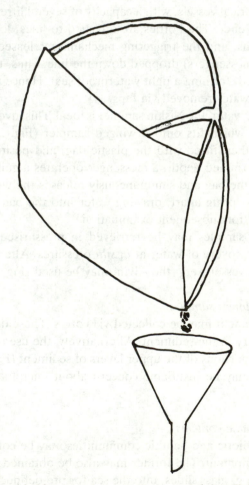

Fig. 2.2 A hand held net used to sample neuston (after Sieburth, 1979).
This instrument is dragged along the uppermost layer of the water column
(preferably on a calm day) and the neuston collected in a bottle.

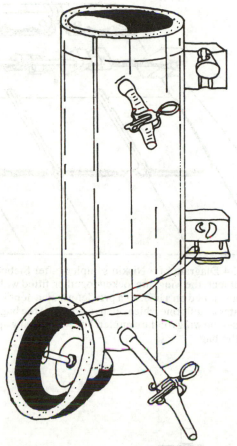

Fig. 2.3 Schematic representation of a Nansen bottle, with the bottom and top sealed and open, respectively. The bottle, with both ends open, is lowered on a line to a pre-determined depth whereupon the top and bottom rubber seals are closed following a triggering mechanism, which is activated by means of a messenger weight. The Nansen bottle is a standard oceanographic instrument, which is used for the collection of non-sterile water samples.

are useful for the recovery of plankton. Essentially, fractions of different sizes may be obtained by using nets of different mesh sizes.

2.2   Microscopic examination of the samples

Microscopic examination of fresh material within minutes of its collection or later, if preserved with formalin or glutaraldehyde, provides useful information about the complexity of the microbial community. Light microscopy (notably phase contrast microscopy) is particularly helpful in

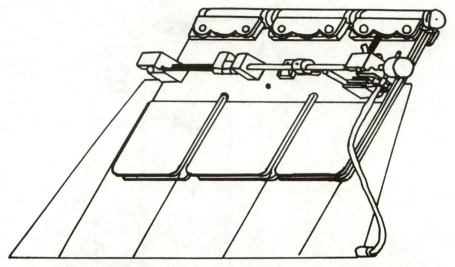

Fig. 2.4 Diagram of a Niskin sampler (after Sieburth, 1979). With this
instrument, the wings are locked together, fitted with a sterile plastic bag,
and positioned on a line. At a pre-determined depth, a messenger weight
operates a guillotine which opens a hose into the bag and simultaneously
releases the wings of the sampler. The wings spring apart, drawing water
into the bag.

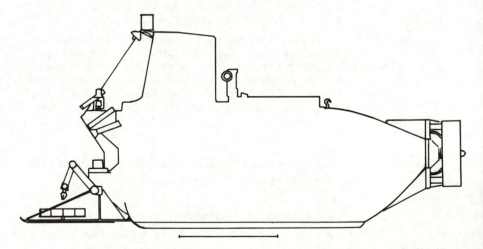

Fig. 2.5 Schematic diagram of the deep-sea submersible Alvin (after
Austin, 1987). This vessel, which is just large enough to accommodate
two-three personnel, is capable of operating at the bottom of the deep sea.

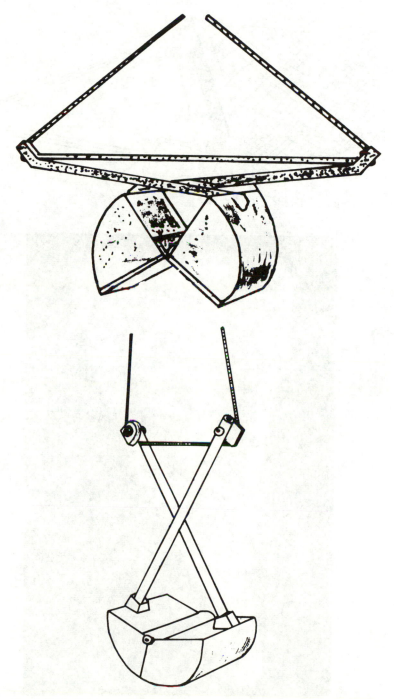

Fig. 2.6 Open (a) and closed (b) grabs used for the recovery of sediment.

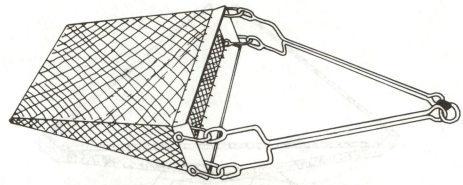

Fig. 2.7 Dredge.

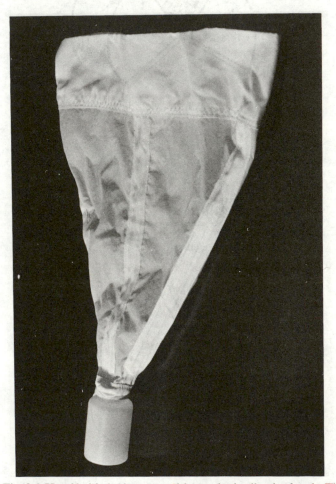

Fig. 2.8 Hand held plankton net with attached collecting bottle. This net was designed to operate from a wooden pole.

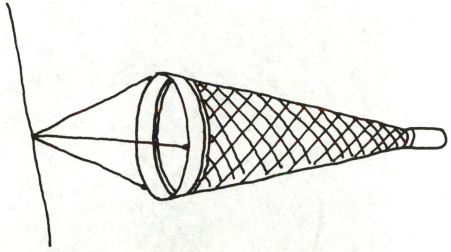

Fig. 2.9 Shipboard plankton net as used on a line.

assessing the comparative abundance of algae, bacteria, fungi and protozoa, such as in biofouling (epibiotic) communities (Fig. 2.10). An indication of the relative activity of components of the microflora may be gained by use of epifluorescence microscopy on specimens stained with a fluorochrome, e.g. acridine orange. This technique will be considered further in Chapter 5. However, these methods are of negligible use in identifying components of the bacterial microflora. Nevertheless, it is possible to visualise certain micro-organisms, i.e. bacteria, by use of fluorescent antibody techniques. Here, there is an obvious requirement for specific antisera, which are sadly lacking for most marine bacteria. Unfortunately, this reflects the incomplete knowledge of the taxonomy and serological inter-relationships of such organisms. Finally, it is worthy of mention that scanning electron microscopy has value in the study of microbial communities, such as colonisers of exposed surfaces, e.g. dockyard pilings (Fig. 2.11) or in enclosed environments, such as the gastro-intestinal tract of animals.

Although microscopic methods provide useful information about the range of micro-organisms present in any given habitat, it is necessary to resort to the other approaches to accurately determine the amount of microbial material. Such methods include determination of biomass, i.e. the mass of living organisms present in a given habitat, and total viable counts.

2.3   Determination of biomass

It is indeed fashionable in many laboratories to expend much effort in estimating microbial biomass in terms of specific chemicals which occur

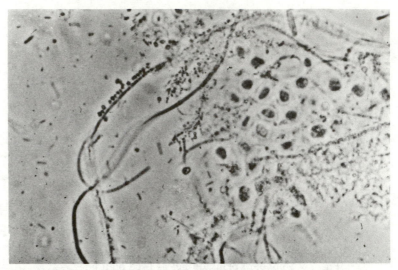

Fig. 2.10 Phase-contrast photomicrograph of a biofouling community, comprising bacteria attached to fungal hyphae, diatoms and other unicellular algae. The sample consisted of a visible layer of slime at the air–water interface of a tank, which contained turbot (from Austin, 1983).

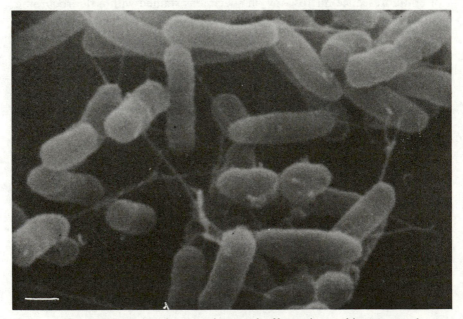

Fig. 2.11 Scanning electron micrograph of bacteria attaching to a wooden surface immersed in the marine environment. Note the presence of apparent attachment fibres. Bar equals 1μm.

in micro-organisms. Such chemicals should, of course, be restricted to living cells, and should not occur elsewhere in the abiotic marine environment. Methods are available for the measurement of:
- *adenosine triphosphate (ATP)* and *DNA*, which are useful for the estimation of the size of the complete microbial community,
- *chitin*, which is present in fungal cell walls,
- *phospholipids*, which are present in prokaryote and eukaryote cell membranes,
- *chlorophyll*, which occurs in algae, *Prochloron* and cyanobacteria, and
- *bacteriochlorophyll, lipopolysaccharide (LPS)* and *muramic acid*, which are constituents of many marine bacteria.

Of these compounds, interest has focused particularly on the measurement of ATP and LPS. This former assay is based upon the luciferin–luciferase reaction, which is responsible for the light emitted by the firefly. In essence, the amount of light produced in the reaction is linearly proportional to the quantity of ATP present in the sample. An excellent overview on the technique has been published by Karl (1980). After filtration to remove organisms of undesired size ranges, ATP is extracted from the environmental sample in boiling Tris (tris-hydroxymethyl-aminomethane hydrochloride) buffer (alternatively acids or solvents may be used) and added to a luciferin–luciferase mixture contained within a tube placed in a dark chamber. Then the production of light is assessed by photomultipliers, amplifiers and recorders. According to Seliger and McElroy (1960), one quantum unit of light is derived from each molecule of ATP hydrolysed in the reaction:

$$\text{ATP} + \text{luciferin(reduced)} + O_2 \xrightarrow{\text{luciferase}} \text{AMP} + \text{pyrophosphate} + CO_2 + \text{product} + \text{light}$$

From data obtained with marine bacteria (Hamilton and Holm-Hansen, 1967), which are similar to data on the algae (e.g. Holm-Hansen and Booth, 1966), it would appear that the ATP content of so-called normal sized bacteria (i.e. the size of typical cells present in colonies growing on nutrient-rich laboratory media) is between 0.5×10^{-9} and $6.5 \times 10^{-9} \mu g$ ATP/cell. This amount is equivalent to approximately 0.3–1.1% of the cell carbon (Hamilton and Holm-Hansen, 1967). Although such information may be used to equate to the number of cells present in a given sample, it should be emphasised that significant errors will be introduced insofar as marine bacteria occur in a wide range of sizes from very small, e.g. some Gram-negative cells, to very large e.g. actinomycetes. In the case of the former, there are marine bacteria which are capable of passing through the 0.22 μm pores of bacteriological filters. Such a wide range of bacterial sizes will undoubtedly introduce an error component in any subsequent extrapolation of the quantity of ATP to microbial population density.

The measurement of LPS, which is present in the cell walls of Gram-negative bacteria, involves use of a purified lysate (aqueous extract) of the blood cells (amoebocytes) of the horseshoe crab (*Limulus polypherus*) (Sullivan and Watson, 1974). This assay is referred to as the Limulus amoebocyte lysate assay (LAL). With this method, at an incubation temperature of 37 °C, the purified amoebocyte lysate reacts specifically with LPS to produce a turbid suspension (at low concentrations of LPS) or a clot (at higher concentrations). Unfortunately, the usefulness of the assay is marred by a scrupulous need for prevention of contamination (Watson and Hobbie, 1979), which reflects the widespread presence of bacteria and, thus, LPS. For example, any reusable glassware and chemicals need to be heated at 180 °C for three hours to ensure the elimination of residual LPS. Moreover, there is an absolute requirement for LPS-devoid distilled water. After extracting samples by boiling for one minute, the amount of LPS is determined by mixing with a pre-determined quantity of amoebocyte lysate. The presence of clots or turbidity may be assessed visually or spectrophotometrically at 360 nm. The amount of LPS may be equated to bacterial population densities using a conversion factor of 2.78 ±1.42 femtograms (fg) of LPS/bacterial cell (Watson *et al.*, 1977). However, this assumes that the quantity of LPS in bacterial cell walls is fairly constant (the experimental evidence for this assumption is lacking). Secondly, as LPS does not occur in Gram-positive bacteria, an estimate of bacterial numbers will reflect only Gram-negative cells.

Some interest has centred on phospholipids, which comprise approximately 50% and 98% of eukaryotic and prokaryotic membrane lipids, respectively (White, 1983). In sediments, the phospholipid content appears to be highly correlated with the amount of extractable ATP (White *et al.*, 1979). For the assay, the lipids are extracted with chloroform and methanol, digested with perchloric acid, and the released phosphate measured by colorimetry.

Measurement of chlorophyll and bacteriochlorophyll is well documented (Parsons *et al.*, 1984; Stal *et al.*, 1984). The chlorophylls must be extracted with methanol or acetone either directly from samples, or after concentration on filters. The quantity of pigment is then determined spectrophotometrically. The amount of chlorophyll may be estimated from the formula:

$$\text{chlorophyll }\mu g/l = \frac{V_s E \times 10^3}{V_i A I}$$

where V_s corresponds to volume (ml) of solvent, E is absorbance in solvent, V_i volume of sample, A is absorption coefficient (g/l/cm), and I is light path of cuvette (cm). However, it should be emphasised that the quantity of chlorophyll contained within cells is dependent upon the physiological state.

2.4 Culturing techniques

A wealth of methods is available for recovery of viable micro-organisms from the marine environment by means of culturing techniques. However there is not any method which will permit the recovery of all microbial cells. Therefore, meaningful data result from the use of a wide range of techniques, each of which permits the study of a small fraction of the total marine microbial biomass. Attention should be given to the need for non-selective or selective methods, to incubation temperatures, and atmospheric conditions, e.g. aerobic, microaerophilic or anaerobic. Clearly, unlike their freshwater or terrestrial counterparts, there may be a requirement for special conditions, such as sodium chloride or the use of pressure chambers to recover deep-sea (barophilic) organisms.

2.4.1 *Prokaryotes*

It may be argued that there have been substantive improvements in cultivation techniques of prokaryotes, notably for specialised groups of bacteria such as cyanobacteria, methanogens and sulphate-reducers, since the pioneering work of Zobell (Zobell, 1941, 1946). Of course, Zobell's methods have stood the test of time, as illustrated by the widespread use of his 2216E agar for the isolation and cultivation of aerobic heterotrophs. Starting soon after (i.e. within an hour) the acquisition of samples, collected in an aseptic manner, scientists will inevitably use agar or broth techniques to cultivate various components of the microflora. The samples may need to be homogenised and/or diluted, for which artificial seawater mixes (Table 2.1) or the simple marine salts solution (Table 2.2) should suffice. Aliquots may be introduced onto non-selective nutrient-rich agar, e.g. Zobell's 2216E medium (Table 2.3). In addition, Simidu's medium (Table 2.4) has fared well, in this author's laboratory, for the recovery of aerobic heterotrophic bacteria. Alternatively, or additionally, nutrient-limited media such as sea-water gelled with 1.5% (w/v) agar, may be used to reveal the presence of bacteria, i.e. oligotrophs, which are inhibited by the presence of high concentrations of organic nutrients. Incubation of low-nutrient media in light may additionally reveal the presence of phototrophs and chemolithotrophs.

Recovery of selected representatives of the marine microflora result from use of specialised media or techniques. Examples include the media for sulphate-reducing bacteria (Table 2.5), phototrophs (Table 2.6), luminous bacteria (Table 2.7) and vibrios (Table 2.8). Selective enrichment for phototrophs may take place in Winogradski columns, in which the concentration of sulphide, vitamins and nutrients, pH, light intensity and temperature

Table 2.1. *Formulation of artificial seawater*

Component	Proportion (% w/v)
Calcium chloride dihydrate	0.147
Boric acid	0.0026
Potassium chloride	0.068
Magnesium chloride dihydrate	1.078
Sodium chloride	2.35
Ethylenediaminetetra-acetic acid tetrasodium salt[a]	0.00003
Sodium hydrogen carbonate	0.0196
Sodium silicate[a]	0.003
Sodium sulphate	0.4
Experimental conditions; pH 8.0	

[a]Used for cultivation of algae.
After Kester *et al.* (1967).

Table 2.2 *Marine salts solution; a suitable diluent for marine bacteria and fungi*

Component	Proportion (% w/v)
Magnesium sulphate (hydrated)	0.7
Potassium chloride	0.075
Sodium chloride	2.4 (higher concentrations will precipitate out during autoclaving)
Experimental conditions; pH 7.0; sterilise at 121 °C/15 min, small glass (ballotini) beads may be added to facilitate agitation	

After Austin *et al.* (1979).

Table 2.3 *Formulation of Zobell's 2216E medium, for the growth of aerobic, heterotrophic marine bacteria*

Component	Proportion (% w/v)
Bacteriological peptone	0.5
Yeast extract	0.1
Ammonium nitrate	0.00016
Boric acid	0.0022
Calcium chloride	0.18
Disodium hydrogen phosphate	0.0008
Ferric citrate	0.01
Magnesium chloride	0.88
Potassium bromide	0.008
Potassium chloride	0.055
Sodium bicarbonate	0.016
Sodium chloride	1.945

Table 2.3 *(cont.)*

Component	Proportion (% w/v)
Sodium fluoride	0.00024
Sodium silicate	0.0004
Sodium sulphate	0.0324
Strontium chloride	0.0034
Agar	1.5
Experimental conditions; pH 7.6; sterilise at 121 °C/15 min	

Based on Zobell (1941).

Table 2.4 *Simidu's medium, which is suitable for the recovery of aerobic heterotrophic bacteria*

Component	Proportion (% w/v)
Bacteriological peptone	0.04
Lab-lemco	0.01
Yeast extract	0.02
Calcium lactate	0.068
Mannitol	0.03
Sodium malate	0.046
Sucrose	0.03
Calcium glycenophosphate	0.01
Ferric citrate	0.001
Ammonium sulphate	0.06
Sodium nitrate	0.06
Tween 80	0.0025
TES[a] buffer	0.115
Agar	1.5
Distilled water	100 ml
Artificial seawater[b]	900 ml
Experimental conditions; pH 7.8	

[a]TES = N-tris (hydroxymethyl) methyl-2-aminoethane sulphonic acid.
[b]Artificial seawater contained only 25% of the normal levels of divalent cations, i.e. Ca^{2+}, Mg^{2+} and Sr^{2+}, in natural seawater. However, the mix was supplemented with 300 μg/l sodium silicate; 170 μg/l lithium sulphate; 60 μg/l potassium iodide; 60 μg/l barium chloride; 40 μg/l indium sulphate; 20 μg/l potassium aluminium sulphate; 50 μg/l zinc chloride; 50 μg/l sodium molybdate; 40 μg/l selenous acid; 30 μg/l copper sulphate; 30 μg/l stannous chloride; 20 μg/l manganous chloride; 20 μg/l nickel sulphate; 20 μg/l vanadyl chloride; 10 μg/l titanium chloride; 10 μg/l cobaltous sulphate; 3 μg/l potassium chromate.

After Simidu (1974).

Table 2.5 *Medium for the recovery of sulphate-reducing bacteria*

Component	Composition (% w/v)
Medium I	
Beef extract	0.1
Calcium chloride	0.01
Dipotassium hydrogen phosphate	0.05
Ferrous ammonium sulphate	0.0392
Magnesium sulphate	0.2
Peptone, bacteriological	0.2
Sodium ascorbate	0.01
Sodium lactate	0.35
Sodium sulphate	0.15
Experimental conditions; pH 7.5; sterilise at 121 °C/15 min	
Medium II	
Ammonium sulphate	0.01
Calcium chloride	0.01
Dipotassium hydrogen phosphate	0.2
Ferric chloride	0.002
Magnesium sulphate	0.01
Sodium thiosulphate	1.0
Experimental conditions; pH 7.8; sterilise at 121 °C/15 min	

After Anon (1975).

Table 2.6 *'Pfennigs medium' for the growth of photo-trophs*

Component	Quantity	
Trace Elements Solution		
Cobaltous chloride hexahydrate	190	mg
Cupric chloride	17	mg
Ethylenediaminetetra-acetic acid disodium salt	5.2	g
Boric acid	62	mg
Ferrous chloride	1500	mg
Manganous chloride	100	mg
Nickel chloride	24	mg
Sodium molybdate	36	mg
Zinc chloride	70	mg
Distilled water	1	l
Solution A		
Calcium chloride dihydrate	0.05	g
Potassium dihydrogen orthophosphate	1	g
Magnesium sulphate heptahydrate	3	g

Table 2.6 *(cont.)*

Component	Quantity	
Ammonium chloride	0.5	g
Sodium chloride	20	g
Trace elements solution	1	ml
Distilled water	950	ml
Solution B		
Vitamin B$_{12}$	2	mg
Distilled water	950	ml
Solution C		
Sodium hydrogen carbonate	50	g
Distilled water	1	l
Solution D		
Sodium sulphide nonahydrate	50	g
Distilled water	1	l
Feeding solution		
Sodium sulphide nonahydrate	30	g
Distilled water	1	l

Equal volumes of solutions A to D are mixed, and used to completely fill 50 or 100 ml capacity screw cap bottles. The medium should be adjusted to pH 6.8 or 7.2 for green sulphur bacteria or purple sulphur bacteria, respectively.

After Trüper (1970); Biebl and Pfennig (1978); Pfennig and Trüper (1981).

Table 2.7 *Isolation medium for luminescent Photobacterium spp.*

Component	Proportion (% w/v or v/v)
Glycerol	0.3
Bacteriological peptone	0.5
Yeast extract	0.3
Agar	1.5
Aged seawater	75
Distilled water	25
Experimental conditions; pH 7.8	

After Schneider and Rheinheimer (1987).

assumes importance (Imhoff, 1987). Methanogens may be recovered in the roll tube method (Hungate, 1969) and in agar deeps/shake tubes, incubated anaerobically (Conrad and Schütz, 1987). Damaged cells may be recovered by pre-incubation of samples in broth before transfer to solid media (see

Olson, 1978). Liquid media are effective for the recovery of many organisms, enrichment of which may be by most probable number techniques. For example, solutions of lactose broth (Table 2.9) and 1% (w/v) peptone at pH 8.4–8.5 are suitable for the development of coliforms (these are lactose-fermenting asporogenous Gram-negative rods) and vibrios, respectively. Nevertheless, it must be emphasised that it is often exceedingly difficult to distinguish true resident marine bacteria from chance contaminants or transient populations.

After inoculation of media, the dilemma continues concerning the nature and extent of the incubation conditions. Psychrophiles, such as may be

Table 2.8 *Thiosulphate citrate bile salt sucrose agar; an effective selective isolation medium for marine vibrios*

Component	Proportion (% w/v)
Bacteriological peptone	1.0
Brom-thymol blue	0.004
Ferric citrate	0.1
Ox bile	0.8
Sodium chloride	1.0
Sodium citrate	1.0
Sodium thiosulphate	1.0
Sucrose	2.0
Thymol blue	0.004
Yeast extract	0.5
Agar No. 1	1.4

Experimental conditions; pH 8.6; autoclaving is not required

Developed by Kobayashi *et al.* (1963).

Table 2.9 *Formula of lactose broth as used for the enumeration of coliforms*

Component	Composition (% w/v)
Bacteriological peptone	0.5
Beef extract	0.3
Lactose	0.5

Experimental conditions; pH 6.8–7.0; sterilise at 121 °C/15 min

After Anon (1970).

expected to occur in the colder polar climates, will require low temperatures, such as 4 °C, whereas mesophiles, found in the tropics, may grow best at 25–30 °C. Moreover, timing of incubation is crucial insofar as some marine bacteria e.g. vibrios, produce colonies within a few days at 25 °C, while others, such as psychrophiles, may require several weeks to produce visible colonies. To date, the majority of work has highlighted aerobic communities, although there is much recent interest in anaerobes, e.g. methanogens. Obviously, recovery of these organisms necessitates the use of anaerobic conditions. Then, to study deep-sea communities, it will be essential to use *in situ* pressures and temperatures. All these factors must be considered in order to achieve a balanced view of marine microbial prokaryotic communities.

2.4.2 *Eukaryotes*

The complexity of populations of marine eukaryotes is reflected in the nature of culturing methods. Although compared to bacteria, marine fungi have been neglected, cultivation techniques have been described. By eliminating bacterial contamination with 0.02% (w/v) chloramphenicol, marine fungi may be recovered in cornmeal agar prepared in seawater. Moreover, media containing wood pulp or cellulose have been successful for cultivation of ascomycetes.

Marine micro-algae (phytoplankton), e.g. *Skeletonema, Dunaliella, Isochrysis* and *Monochrysis*, may be cultured in a largely inorganic cocktail supplemented with vitamins (Table 2.10). These may be used as biomass

Table 2.10 *Media suitable for the cultivation of marine micro-algae*

Component	Concentration (/l)	
Biotin	0.05	μg
Cupric chloride anhydrous	0.2	μg
Ferric chloride anhydrous	72.6	μg
Di-potassium hydrogen phosphate	1.05	mg
Manganese chloride anhydrous	2.3	μg
Ethylenediaminetetra-acetic acid disodium salt	300	μg
Sodium molybdate	2.5	μg
Sodium nitrate	25	mg
Thiamine hydrochloride	0.1	mg
Vitamin B_{12}	0.5	μg
Zinc chloride	2.1	μg

Prepared in artificial seawater (see Table 2.1)
Experimental conditions; sterilise at 60 °C/4 h

From Anon (1975).

for subsequent experimentation, or as food for zooplankton, e.g. copepods. To retard bacterial multiplication, it may be necessary to supplement the media with antibiotics, e.g. oxytetracycline at 15 mg/l.

Zooplankton such as ciliated protozoa, e.g *Tetrahymena*, may be grown in axenic culture at 26 °C in suitable liquid media (Table 2.11). Copepods, e.g. *Acartia*, may be cultured in seawater supplemented with algae, as discussed previously.

Table 2.11 *Medium suitable for the cultivation of ciliated protozoa*

Component	Concentration (/l)	
Proteose peptone	20	g
Yeast extract	2	g
Glucose	5	g
Fe: EDTA chelate	90	μg
Prepared in artificial seawater (see Table 2.1) Experimental conditions; sterilise at 121 °C/15 min		

Anon (1975).

3   Quantification of marine microbial populations

Of course, the numbers of micro-organisms, and indeed their biomass, recorded from the marine environment will reflect the nature of the methods used (see Chapter 2). However, by way of generalisation, it would appear that the largest populations are present in the uppermost layer of water (neuston) and sediments in warm, i.e. tropical and sub-tropical locations, which contain abundant organic material. Moreover, there is a seasonal influence coinciding with changes of temperature, with maximal and minimal numbers often present during summer and winter, respectively. Yet the marine microbial populations should not be construed as exhibiting homogeneous distributions, insofar as rapid responses to localised changes, such as the effects of tides in coastal sites (Erkenbrecher and Stevenson, 1975), may be apparent during intervals as short as a few minutes. Thus, very regular sampling would be necessary to achieve meaningful data about population dynamics of marine micro-organisms. Marked differences in microbial population sizes may be apparent over very small distances (McAlice, 1970; Krumbein, 1971; Ashby and Rhodes-Roberts, 1976). There is a general tendency to make sweeping generalisations after examination of a comparatively few plate counts or biomass determinations. Such data will not accurately reflect the complex nature of a continually changing microbial population. If anything, most reports, particularly those based on total viable counts, greatly underestimate the size of the marine microflora. Unfortunately, it is a rare occurrence indeed for a research publication to consider more than one type of micro-organism. For example, bacteriologists may well neglect eukaryotes and, for that matter, some of the less familiar forms of prokaryotes. Numerous articles have considered numerical changes in populations of selected groups, e.g. the ever-popular aerobic heterotrophic bacteria (or rather those organisms capable of producing colonies on a single nutrient-rich medium with incubation conditions not necessarily appropriate for the habitat under study) while ignoring other organisms, such as anaerobes and/or chemolithotrophs. Nevertheless by piecing together various sources of information, an overall picture emerges.

3.1   Prokaryotes

3.1.1   *Numbers of bacteria in the marine environment*

As an over-simplification, bacterial populations in seawater usually range between 10^3 and 10^6/ml, with counts of up to 10^9/g recorded for sediments. Overall, there is a gradual decline in bacterial numbers down the water column (Table 3.1), albeit with a slight increase at the thermocline (e.g. Ezura *et al.*, 1974). Maximum populations of up to 10^8/ml are recovered from the neuston (Table 3.2). Certainly, populations peak closest to shore, probably reflecting a greater abundance of organic nutrients. For example, Gunn *et al.* (1982) reported a reduction in the numbers of aerobic heterotrophic bacteria from 10^4/l at a coastal site on the Atlantic seaboard of the USA to 2×10^3/l at an oceanic site. However, these populations, particularly for the coastal site, seem low compared to other workers. Similarly, Jannasch and Jones (1959) and Austin (unpublished data) have noted the comparative dearth of bacteria at sampling sites located at considerable distances from shore.

Data have been published which point to a seasonal distribution in bacterial numbers coinciding with changes in the water temperature. At a coastal site in England, Austin (1983) recovered minimal numbers of aerobic heterotrophic bacteria in January (10^4/ml; water temperature = 3 °C) and maximal populations of 4×10^9/ml (water temperature = 19 °C) in August. Parallel results were reported by Ezura *et al.* (1974) and Kaneko *et al.* (1978) for samples recovered from Japan and the Beaufort Sea (Arctic),

Table 3.1 *Bacterial populations in water of the South China Sea, as determined by direct counts, direct viable counts, i.e. involving use of nalidixic acid, and by a spread plate technique*

Depth at which sample was obtained (m)	Total microscopic count (/ml)	Direct viable count (/ml)	No. of aerobic heterotrophic bacteria by spread plating (/ml)
Station 1			
0	1.2×10^6	8.2×10^4	1.2×10^2
100	6.9×10^5	5.4×10^3	5.5×10^1
600	2.0×10^5	2.4×10^3	8.0×10^0
Station 2			
0	9.1×10^5	4.2×10^4	2.1×10^2
100	6.5×10^5	2.1×10^4	5.3×10^1
600	2.6×10^5	2.4×10^3	1.6×10^1

From Simidu *et al.* (1983).

Table 3.2 *Estimates of the size of bacterial populations in the marine environment*

Source	Number of bacteria	Method of examination	Reference
Neuston			
North Atlantic	>2 × 10^7/ml	Total viable count	Sieburth *et al.* (1976)
North Atlantic	10^5–10^8/ml (10^1–10^4/cm^2)	Filtration (nucleopore filters)	Crow *et al.* (1975)
North Atlantic	10^8/ml	ATP analysis	Sieburth (1979)
Water			
Surface	5 × 10^5– 6 × 10^6/ml	Total viable count	Carlucci (1974)
Surface, thermocline	10^3–10^5/ml	Total viable count	Sieburth (1971), Sieburth *et al.* (1976)
Surface, Swedish coast	1.4 × 10^4–13.5 × 10^4/ml	Total viable count	Kjelleberg and Håkansson (1977)
Surface (0 m depth), Far East	~2.1 × 10^5–2.7 × 10^5/ml	Total viable count (spread plate technique)	Simidu *et al.* (1983)
Surface, over coral reef, tropical	1 × 10^5–2.5 × 10^6/ml	Direct count (epifluorescence)	Moriarty *et al.* (1985)
Surface, Beaufort Sea (Canadian)	10^3–10^4/ml	Total viable count	Bunch and Harland (1976)
Surface, Arctic Ocean (near Greenland)	<1/ml	Total viable count	Kriss (1963)
Surface, Beaufort Sea	1.8 × 10^5–8.2 × 10^5/ml	Direct count	Kaneko *et al.* (1978)
Surface, South China Sea and West Pacific Ocean	1.3 × 10^2–2.5 × 10^3/ml	^14C most probable numbers technique	Ishida *et al.* (1986)
Surface, South Pacific Ocean (Antarctic)	1.3 × 10^5–7.2 × 10^5/ml	Direct count	Hanson *et al.* (1983)
Surface, Atlantic Ocean (oceanic)	10–100/ml	Total viable count	Austin (unpublished data)
Surface, coastal England	1 × 10^4–4 × 10^8/ml	Total viable count	Austin (1983)
Coastal, pelagic, Limfjorden, Denmark	5 × 10^5–15.2 × 10^6/ml	Direct count (epifluorescence)	Andersen and Sørensen (1986)
Coastal, northern Europe (overlying beach sand)	1.2 × 10^6/ml	Total viable count	Meyer-Reil *et al.* (1978)
Coastal, Atlantic Ocean (USA)	10^4/l	Total viable count	Gunn et al. (1982)
Coastal, Atlantic Ocean (USA)	6.6 × 10^5/ml	Direct count (epifluorescence)	Ferguson and Rublee (1976)
Sub-surface, euphotic zone	10^5–10^6/ml	ATP analysis	Sieburth (1979)
Euphotic zone	10^4/ml	ATP analysis	Sieburth (1979)
Sargasso Sea	2.9 × 10^5–3.2 × 10^5/ml	Direct count (epifluorescence)	Caron *et al.* (1982)
Southwest Africa	1.5 × 10^4 – 6.29 × 10^6/ml	Direct count	Watson *et al.* (1977)
Finland	5.5 × 10^5– 4.7 × 10^6/ml	Total viable count	Väätänen (1980)
Norwegian fjord	~10^6/ml	Direct count (epifluorescence)	Dahle and Laake (1982)
Japan	4.0 × 10^3– 4.8 × 10^5/ml	Total viable count	Araki and Kitamikado (1978)
10 m depth, Far East	3.1 × 10^4– 6.9 × 10^4/ml	Total viable count	Simidu *et al.* (1983)
Oceanic	scant population	Total viable count	Jannasch and Jones (1959)
Atlantic Ocean (oceanic)	2 × 10^3/l	Total viable count	Gunn *et al.* (1982)

Table 3.2 *(cont.)*

Source	Number of bacteria	Method of examination	Reference
Suspended particulates			
Plankton, Far East	10^7–10^9/ml	Total viable count	Taga and Matsuda (1974)
Marine snow, Sargasso Sea	6.6×10^5–32.79×10^6/ml	Direct count (epifluorescence)	Caron *et al.* (1982)
Sediment			
Saltmarsh, USA	1×10^6–1.9×10^7/ml	Direct count (epifluorescence, scanning and transmission electron microscopy)	Wilson and Stevenson (1980)
Beach sand, Japan	10^4–10^6/g	Total viable count	Maeda and Taga (1973)
Beach sand, eastern Mediterranean	7.8×10^2–4×10^4/g	Total viable count	Khiyama and Makemason (1973)
Beach sediment, northern Europe	1.3×10^9/g dry wt	Total viable count	Meyer-Reil *et al.* (1978)
Beach sand, Germany	1.7754×10^9/cm^3 wet wt	Direct count (epifluorescence)	Weise and Rheinheimer (1978)
Coastal sediment, Japan	3.9×10^3–5.6×10^5/g	Total viable count	Maeda and Taga (1973)
Marine, Japan	$\sim 10^5$/g	Total viable count	Sugahara *et al.* (1974)
Marine, Japan	3×10^3–2.8×10^6/g	Total viable count	Araki and Kitamikado (1978)
Marine, Beaufort Sea	6.2×10^7–2.1×10^9/g	Direct count	Kaneko *et al.* (1978)
Marine, Bahama Bank	3.4×10^3–3.18×10^6/g wet wt	Total viable count	McCallum (1970)
Sub-surface, Irish Sea	<10 – 6.3×10^5/g dry wt	Total viable count	Litchfield and Floodgate (1975)
Marine, Canada	5.2×10^{10}/g dry wt	Direct count (epifluorescence)	Velji and Albright (1986)
Saltmarsh, USA	6.8×10^8–5.1×10^{10}/g	Direct count (epifluorescence)	Yamamoto and Lopez (1985)

respectively. However it is also conceded that this trend may be reversed (Ezura *et al.*, 1974; Austin, unpublished data).

Apart from temperature, the apparent fluctuation in populations may reflect complex nutritional and physico-chemical variations within the ecological niches. It is recognised that bacterial populations may be considerably reduced by interactions with other micro-organisms. For example, the deleterious effects of zooplankton, i.e. microflagellates and ciliates, phytoplankton, due to grazing by diatoms and other small unidentified organisms and perhaps by the production of extracellular compounds have been well established (Fogg, 1966; Kogure *et al.*, 1980a; Martin and Bianchi, 1980; Davis and Sieburth, 1984; Fuhrman and McManus, 1984; Montagna, 1984; Andersen and Sørensen, 1986; Andersson *et al.*, 1986; Rassoulzadean and Sheldon, 1986). In coastal waters, Wright and Coffin (1984) estimated daily grazing rates of bacteria by zooplankton amounting to approximately 10^6 bacterial cells/ml. This was 33–50% elimination of the standing population of 2×10^6–3×10^6 bacteria/ml. Davis and Sieburth (1984) considered that when bacterial numbers are in the range of 10^5–10^7/ml, grazing by flagellates clears between 30 and 200 bacteria/flagellate/hour. Conversely, it has been demonstrated that zooplankton (Kogure *et al.*, 1980a) and indeed some diatom blooms (Väätänen, 1980) may enhance bacterial numbers, presumably by the provision of nutrients.

Considering a north–south divide, it is apparent that from some studies only low numbers of aerobic heterotrophic bacteria are thought to occur in Antarctic waters, although there is a marked fluctuation between sites (Weibe and Hendricks, 1974). In a detailed study by Hanson *et al.* (1983) using acridine orange direct counts, a fluctuation in bacterial numbers was found in samples taken from the Sub-tropical Convergence Zone to the Polar Front. In surface waters of the eastern South Pacific Ocean, bacterial populations declined from 7.2×10^5/ml in the Sub-tropical Convergence Zone, to 1.3×10^5/ml at the Sub-antarctic Front, but increased again to 3.5×10^5/ml at the Polar Front. Nevertheless, there were only half of the bacterial numbers at the Polar Front compared to the Sub-tropical Convergence Zone.

According to Kaneko *et al.* (1978), higher bacterial numbers occur in the Arctic than Antarctic. However, conflicting data are available. For example, Kriss (1963) reported less than one bacterium/ml in surface water from the vicinity of Greenland. Bunch and Harland (1976) discussed bacterial populations of 10^3–10^4/ml in surface waters of the Canadian zone of the Beaufort Sea. Yet using direct counts on surface waters from the Beaufort Sea, Kaneko *et al.* (1978) mentioned bacterial populations during a summer–winter–summer period of 8.2×10^5/ml, 1.8×10^5/ml and 5.2×10^5/ml, respectively. Furthermore, it is noteworthy that the populations were not significantly less in sea ice compared to the underlying (liquid) water.

It is thought likely that many bacteria (up to 75%) in the water column are located on particulates (Wilson and Stevenson, 1980). Apart from the evidence concerning colonisation of surfaces in the marine environment, which will be discussed later, Taga and Matsuda (1974) calculated 10^7–10^9 bacteria/ml of plankton. More recently, Caron *et al.* (1982) examined marine snow from the Sargasso Sea. Using epifluorescence microscopy, they calculated the bacterial populations as 6.6×10^5–3.279×10^7/ml. Lower populations of 2.9×10^5–3.2×10^5 bacteria/ml existed in the surrounding seawater (Caron *et al.*, 1982). Subsequently, Alldredge *et al.* (1986) estimated that 0.1–4.0% of seawater bacteria occurred in association with marine snow (i.e. dissolved and colloidal organic matter which aggregates together as amorphous suspended masses). Other workers have shown that many micro-organisms are free-living in the sea. For example, Kogure *et al.*

(1980b) determined microscopically that the majority of the organisms in Tokyo Bay were free-living. Indeed, only rarely were particles seen to be heavily colonised.

The data pertaining to bacterial populations on sediments have been summarised in Table 3.1. Essentially, a tremendous range of values has been reported, although, generally <1% of the total surface area of sediment is populated with micro-organisms (Yamamoto and Lopez, 1985). Interestingly, these authors did not find any significant difference in bacterial populations within rich organic muds or coarse sands. Evidence supports the notion of denser populations in the sediment than the overlying seawater. Meyer-Reil *et al.* (1978) estimated the difference in values to be on the order of a factor of 1000. Thus with actual numbers fluctuating considerably, the average number of bacteria in beach sediments of northern Europe was determined as 1.3×10^9/g dry weight. The corresponding figure for the overlying water was 1.2×10^6/ml. Of course, the basis of choosing a meaningful unit quantity could be duly questioned, insofar as the relationship of a millilitre of water to a gram dry weight of sediment is difficult to comprehend. Nevertheless, parallel findings were also obtained by Kaneko *et al.* (1978). Here, a seasonal fluctuation in numbers was apparent. An even more dramatic common finding is that bacterial numbers rapidly decline with increasing depth into the sediments (e.g. McCallum, 1970; Litchfield and Floodgate, 1975). In fact, the decline in populations is substantial within only 13 cm of the sediment surface (McCallum, 1970).

Concerning beaches, there is a marked variation in numbers between studies, from estimates of approximately 7.8×10^2/g up to 1.7×10^9/cm^3 (see Table 3.1). Meadows and Anderson (1966) observed micro-colonies consisting of up to 150 cells on sand grains. Smaller micro-colonies of 5–20 cells were noted by Weise and Rheinheimer (1978), with colonisation principally in protected sites such as fissures. In addition, bacteria were seen to be attached to diatoms. Nevertheless, it was calculated that <5% of the surface area was actually colonised by bacteria. From their observations by epifluorescence microscopy, Weise and Rheinheimer (1978) calculated that total bacterial populations of beach sand amounted to 1.7754×10^9/cm^3 wet weight. This overall microflora contained 8.688×10^8 cells which were considered to be attached to particles, with the remainder in the interstitial areas.

3.1.2   *Quantitative data of specific groups of prokaryotes in the marine environment*

It is often difficult to assess the precise number of different bacterial groups in the marine environment from purely quantitative data. Often such information is obtained from use of selective isolation procedures. For example, using rifampicin as a selective agent, Weber and Greenberg (1981) calculated the numbers of spirochaetes in salt marshes to be 10^4–10^6/g wet weight. Using enrichment culture techniques, Tuttle and Jannasch (1972) recovered 10–100 thiobacilli/100 ml of oceanic seawater. Other examples have been included in Table 3.3. As many as 45–98% of the heterotrophic bacteria in

Table 3.3 *Quantitative data for specific groups of prokaryotes in the marine environment*

Taxonomic grouping	No. of cells	Source	Reference
Actinomycetes	1.48–23/cm^3	Sediment, North Sea	Weyland (1969)
Actinomycetes	23/cm^3	Sediment, Atlantic Ocean	Weyland (1969)
Actinomycetes	300–1270/cm^3	Sediment of shallow sea, Japan	Okami and Okazaki (1978)
Bdellovibrio spp.	~200/l	Coastal water, Hawaii	Taylor *et al.* (1974)
Beggiatoa spp.	>2 × 10^4/ml	Sediment	Lackey and Clendinning (1965)
Cocci, Gram-positive	60/100ml	Coastal water, Atlantic Ocean	Gunn *et al.* (1982)
Cocci, Gram-positive	20/100ml	Water of open ocean, Atlantic Ocean	Gunn *et al.* (1982)
Cyanobacteria	5 × 10^5–1.5 × 10^6/ml	Water, Pacific Ocean	Li *et al.* (1983)
Cyanobacteria	10^4–10^5/ml	Coastal water, Atlantic Ocean	Murphy and Haugen (1985)
Cyanobacteria	10^3–10^4/ml	Oceanic water, Atlantic Ocean	Murphy and Haugen (1985)
Cyanobacteria, small coccoid	10^4–10^5/ml	Water, Banda Sea (Indonesia)	Zevenboom (1986)
Cytophaga spp.	1–100/ml	Inshore water, Japan	Kadota (1959)
Cytophaga spp.	10^1–10^4/g	Inshore mud, Japan	Kadota (1959)
Halobacteria	7 × 10^6/ml	Surface water, Dead Sea	Kaplan and Friedmann (1970)
Halobacteria	7 × 10^5/ml	Water–100 m depth, Dead Sea	Kaplan and Friedmann (1970)
Halobacteria	2 × 10^4/ml	Water–200 m depth, Dead Sea	Kaplan and Friedmann (1970)
Luminous bacteria	1 × 10^3–6 × 10^3/l	Coastal water, California	Ruby and Nealson (1978)
Luminous bacteria	1 × 10^1–6.3 × 10^2/l	Water–160–320 m depth, Sargasso Sea	Orndorff and Colwell (1980)
Methanogens	17–9300/g dry wt	Coastal sediment–upper 5 cm, USA	Hines and Buck (1982)
Methanogens	78–39 000/g dry wt	Coastal sediment– 13–18 cm depth, USA	Hines and Buck (1982)
Nitrifying bacteria	10^1–10^3/g	Sediment, Japan	Sugahara *et al.* (1974)
Nitrifying bacteria (oxidises ammonia)	0–36/100ml	Water, Japan	Ezura *et al.* (1974)
(Ammonia producers)	330–1100/ml	Water–0–10 m depth, Japan	Ezura *et al.* (1974)
(Ammonia producers)	330–540/ml	Water–20 m depth, Japan	Ezura *et al.* (1974)
Nitrogen-fixing bacteria	1–10g	Offshore sediment, Japan	Sugahara *et al.* (1974)

Table 3.3 *(cont.)*

Taxonomic grouping	No. of cells	Source	Reference
Phototrophs containing bacterio-chlorophyll *a*	~10^3/g	Beach sand, Japan	Maeda and Taga (1973)
Phototrophs containing bacterio-chlorophyll *a*	4.8×10^2–1.4×10^5/ml	Seawater, Japan	Maeda and Taga (1973)
Spirochaeta spp.	10^4–10^6/g wet wt	Salt marsh	Weber and Greenberg (1981)
Sulphate reducers	130–36 000/g dry wt	Coastal sediment– 2–5 cm depth, USA	Hines and Buck (1982)
Sulphate reducers	0–860/g dry wt	Coastal sediment– 13–18 cm depth, USA	Hines and Buck (1982)
Sulphate reducers	0/g dry wt	Coastal sediment– >23 cm depth, USA	Hines and Buck (1982)
Synechococcus spp.	3×10^5/ml	Water (euphotic zone)	Waterbury and Stanier (1981)
Synechococcus spp.	2×10^5/ml	Water over coral reef	Moriarty *et al.* (1985)
Thiobacillus spp.	0–100/ml	Water, open sea	Tilton *et al.* (1967)
Thiobacillus spp.	10–100/ml	Water, open sea	Tuttle and Jannasch (1972)
Halophilic vibrios	90–6700/ml	Coastal water, Hong Kong	Chan *et al.* (1986)

the South China Sea and the west Pacific Ocean were considered by Ishida *et al.* (1986) to be obligate oligotrophs. Such organisms are unlikely to grow on nutrient-rich media. It is recognised that some taxa dominate in certain habitats. For example, halophilic vibrios may account for 0.4–40% of total bacterial populations in sub-tropical coastal water (Chan *et al.*, 1986). In the presence of hydrocarbon pollutants, *Alcaligenes* may comprise >50% of the total number of bacteria (Westlake *et al.*, 1976). In seawater, *Bacillus* spp. may comprise only 0–16% of the total aerobic heterotrophic bacterial microflora, yet in sediments the proportion may reach 100% (Bonde, 1975; 1976).

Data highlight the comparative sparseness of actinomycetes in the marine environment (e.g. Weyland, 1969; Okami and Okazaki, 1978). Their presence is overwhelmingly restricted to coastal sediments, where populations of between 1×10^3 and 1.27×10^3/cm^3 occur (Weyland, 1969; Okami and Okazaki, 1978). Seemingly, *Micromonospora* constitutes the dominant actinomycete, at least in beach sand (Watson and Williams, 1974). Other Gram-positive organisms, such as micrococci–staphylococci, are also quite rare. In one study, Gunn *et al.* (1982) found only 20–60 cells/100 ml of water. Similarly, the luminous bacteria occur in low numbers, 10^1–6×10^3/l

(Ruby and Nealson, 1978; Orndorff and Colwell, 1980). Orndorff and Colwell (1980) considered that these organisms constituted between 0.6 and 7.6% of the bacterial populations. Apart from their association with fish, luminous bacteria appear to be free-living in the water column (Ruby and Nealson, 1978).

Of the specialised physiological groups, methanogens and sulphate-reducers are more common in sediment than the water column, although the reverse is true for the phototrophs. In an examination of coastal sediments off the USA, Hines and Buck (1982) determined that methanogens were more numerous than sulphate-reducers (see Table 3.3), despite an overall reduction in numbers with increasing depth into the sediment. Methanogens peaked (i.e. up to 3.9×10^4/g dry weight) in sub-surface sediments, at depths of 13–18 cm, and then declined sharply. With sulphate-reducers, there was a steady decline in numbers throughout the sediment layers. Here, maximal populations of 3.6×10^4/g dry weight were found up to 5 cm into the sediment. At depths of greater than 23 cm, these organisms were not recovered. Phototrophs may normally be present in seawater in numbers up to 10^6/ml. Organisms with bacteriochlorophyll *a* were found in population densities of 4.8×10^2–1.4×10^5/ml in seawater off the coast of Japan (Maeda and Taga, 1973). In certain circumstances cyanobacteria, namely *Lyngbya*, *Nostoc* and *Trichodesmium*, form visible blooms on the surface of seawater. Here, the population size is $>10^8$/ml. In other circumstances, population densities of 10^3–10^5/ml occur (Fogg, 1987). Cyanobacteria are common in warm water, particularly in tropical seas (Waterbury *et al.*, 1979). According to Zevenboom (1986) 10^4–10^5 small coccoid cyanobacterial cells/ml were recovered from deeper waters of the Banda Sea off Indonesia, where they accounted for between 10 and 50% of the total phytoplankton community. Cyanobacteria have been reported to comprise 8.3–79.4% of the total picophytoplankton (or 4.7–46.4% of the total phytoplankton) communities in coastal waters off Japan (Takahashi *et al.*, 1985). *Synechococcus* and possibly *Synechocystis* (Murphy and Haugen, 1985) comprise the principal components of photosynthetic biomass in oceanic water (Campbell and Carpenter, 1986), where populations may reach 2×10^5/ml (Moriarty *et al.*, 1985). In temperate waters, there is a seasonal fluctuation in numbers of cyanobacteria (El Hag and Fogg, 1986).

The seasonal fluctuation in numbers of aerobic heterotrophic bacteria at a coastal site in England, reported by Austin (1983), could be explained by responses of selective components of the microflora. In the seawater during winter, stalked (*Hyphomonas–Hyphomicrobium*), yellow-pigmented (*Cytophaga–Flavobacterium–Flexibacter*) photobacteria and vibrios predominated. Additionally during summer, *Acinetobacter, Bacillus, Micrococcus–Staphylococcus* and *Pseudomonas* were recovered. Therefore, it is tempting to infer that the tremendous increases in bacterial numbers between winter and summer were attributable to Gram-positive taxa, and acinetobacters and pseudomonads. Such organisms may well represent terrestrial forms, constituting a transient, wash-in component in the coastal marine environment. Consequently, it seems unlikely that quantitative studies will reliably differentiate truly indigenous marine micro-organisms from transient forms. Novitsky and Karl (1985) noted the influence of sewage outfalls on microbial populations. This is also illustrated by bacterial counts in marine fish farm effluent reaching the sea. In particular, at the point of contact with seawater, Austin (1983) reported that the effluent contained 5×10^5 bacteria/ml in January (temperature = 5 °C), increasing to a massive count of 1.1×10^{10} bacteria/ml in August (temperature = 20 °C). The effects of such high bacterial numbers on receiving waters is subject to conjecture, but such evident sources of pollution undoubtedly contribute to the apparently high bacterial numbers recorded at coastal sites.

3.1.3   *Influence of methods on the estimation of bacterial populations*

In the preceding sections, the population data have been obtained using a variety of methods, most of which may be categorised as total viable counts using agar plates, or direct counts by microscopy. For the uninitiated, it is difficult to ascertain which techniques provide the most meaningful data. In a fascinating study, Velji and Albright (1986) compared the differences in sediment bacterial populations obtained using standard epifluorescence techniques (population = 5.2×10^{10}/g wet weight) and following treatment of samples with pyrophosphate and ultrasound (population = 11.3×10^{11}/g). Great anomalies are apparent between direct and viable counts (Jannasch and Jones, 1959; Kogure *et al.*, 1979; Dahle and Laake, 1982; Simidu *et al.*, 1983; Staley and Konopka, 1985) with differences attributed to the presence of dormant/non-culturable cells, clumping, dead cells and uniformly-shaped inanimate particles which resemble micro-organisms. In fact, the plate count technique may account for <0.1% of the organisms observed microscopically (Kogure *et al.*, 1979; Simidu *et al.*, 1983; Table 3.2). Yet the evidence of direct counts may also be misleading insofar as microscopy often reveals the presence of very small (coccoid) objects in higher numbers than readily recognisable bacteria (e.g. Daley and Hobbie, 1975; Wilson and Stevenson, 1980). The significance of these 'mini-cells' will be discussed further in Chapter 5.

To improve the microscopic techniques then available, Kogure *et al.*

(1979) devised a method which permitted the recognition by microscopy of viable cells. This method incorporated nalidixic acid, which arrested cell division and therefore gave rise to the development of long cells. By microscopy, it may be safely assumed that such cells are indeed viable; thereby comprising a direct viable count. Using this method, Kogure *et al.* (1979) estimated that bacterial numbers in the open ocean were three times higher than data from plating but significantly lower than the more conventional direct count. Data for a subsequent comparison of water samples, obtained from the South China Sea, have been included in Table 3.2. In the highly eutrophic environment of Tokyo Bay, the direct viable count was 1.5–39.8% of the total direct count. Offshore direct viable counts of 0.7–7.9% were recorded (Kogure *et al.*, 1980b).

Another approach has been to confirm direct counts by transmission electron microscopy. In their studies of the Sargasso Sea and water from the coast of southwest Africa, Watson *et al.* (1977) noted that for direct counts of 1.15×10^6 bacteria/ml and 10^6 bacteria/ml obtained by epifluorescence microscopy the corresponding values by transmission electron microscopy were 6.59×10^5/ml and 6.2×10^5/ml, respectively. This suggests that many objects observed by light microscopy may be confused for bacteria, thereby artificially inflating the true value. Moreover, these workers concluded that 21% of the bacteria, as observed by transmission electron microscopy, were mini-cells of only 0.0415 μm^3 average size. This compared with 0.3339 μm^3 for vibrios, which comprised 32% of the total number of bacteria.

The problems of total viable counts have been well voiced. Apart from the choice of media and incubation conditions, the effect of pour plating, spread-plating and filtration may cause substantial variation in results (Table 3.4; Patrick, 1978; Simidu *et al.*, 1983). Effectively, the highest counts are

Table 3.4 *Variation in numbers of aerobic heterotrophic bacteria from New Zealand seawater obtained by means of a pour plate, spread plate and membrane filtration technique*

Technique	No. of aerobic heterotrophic bacteria (/ml)
Pour plate	3.1×10^3–1.6×10^7
Spread plate	2.3×10^3–2.2×10^7
Membrane filtration	2.8×10^3–2.5×10^7

Data from Patrick (1978).

obtained by the spread-plate technique (Simidu *et al.*, 1983). Conceivably, many marine bacteria are heat-sensitive and would be damaged in the molten, albeit cooled, agar used for pour plates. With filtration, it may be assumed that many mini-bacteria pass through the pores, thereby contributing to a sizeable error in population estimates. In addition the high levels of nutrients in most commonly used bacteriological media may well retard the growth of oligotrophs, many of which occur in marine (Yanagita *et al.*, 1978; Ishida *et al.*, 1986) and, for that matter, estuarine environments (Mallory *et al.*, 1977).

3.1.4   *Biomass*

   In addition to estimating the numbers of bacteria in sediment and water, some workers have equated the data to biomass. For example, Meyer-Reil *et al.* (1978) determined that the bacterial counts for water and the underlying sediment of 1.2×10^6/ml and 1.3×10^9/g dry weight were equivalent to a biomass of 1.6×10^{-4} mg/ml and 2.9×10^{-1} mg wet weight/g dry weight, respectively. The bacterial populations of 10^4/ml in plankton (Jannasch and Jones, 1959) were considered to represent a biomass of 4 mg carbon (C)/m^3 (Sorokin, 1971). Watson *et al.* (1977) reported direct counts of 1.5×10^4–6.29×10^6 bacteria/ml, with corresponding LPS amounts of 0.04–17.8 ng/ml. Here, the average concentration of LPS per bacterium represented 2.78 $\pm$1.42 fg, which is almost meaningless in terms of the vast range of bacterial sizes occurring in marine habitats. Maeda and Taga (1979) carried out LPS analyses on samples obtained from Tokyo Bay. Essentially, the total LPS amounted to 1.3–6.1 ng/ml, where the free LPS was 0.3–1.9 ng/ml. They noted that the quantity of LPS declined with increasing depth, thus parallelling the results of plate counts. As for ATP, numerous scientists have carried out analyses on water (Sieburth *et al.*, 1976; Karl, 1978) and sediment (Karl, 1978; Novitsky and Karl, 1985). A maximum of 10^4–10^5 fg ATP/ml was found in the euphotic zone of deep-sea stations. Corresponding cell populations and quantities of carbon were 10^4–10^5/ml and 1.5–8.0 mg C/m^3, respectively (Sieburth, 1979). In the Black Sea, Karl (1978) found that due to the presence of metabolically active cells, ATP concentrations at the O_2–HS interface were 5–10 times higher than in the oxygenated environment at a similar depth. The published data for water and sediment showed values of 10 ng ATP/l of water and 150 ng ATP/cm^3 of sediment.

3.2   Eukaryotes

   Data for many groups of eukaryotic micro-organisms have been included in Tables 3.5 and 3.6. It is noteworthy that detailed quantitative

Table 3.5. *Populations of eukaryotic micro-organisms present in the marine environment*

Type of micro-organism	No. of cells	Source	Reference
Amoebae	$1.2–1.35 \times 10^3/l$	Neuston, North Atlantic	Sieburth *et al.* (1976)
	0–0.31/ml	Water, Sargasso Sea	Caron *et al.* (1982)
	0–22/ml	Marine snow, Sargasso Sea	Caron *et al.* (1982)
Ciliates	0/ml	Water, Sargasso Sea	Caron *et al.* (1982)
	0–23/ml	Marine snow, Sargasso Sea	Caron *et al.* (1982)
	1.4–162/ml	Water, Limfjorden (Denmark)	Andersen and Sørensen (1986)
Dinoflagellates	$10^3/l$	Water, North Pacific	Allen (1941)
	$10^4–10^5/l$	Water, Norwegian fjord	Hasle (1950)
	$10^7/l$	Dinoflagellate bloom	Seliger *et al.* (1970)
Flagellates	0.02–0.61/l	Water, Sargasso Sea	Caron *et al.* (1982)
	3–2400/l	Marine snow, Sargasso Sea	Caron *et al.* (1982)
Phytoflagellates (haptophytes)	$10^3/l$	Water, temperate region	Honjo and Okada (1974)
	$10^5/l$	Water, tropical region	Honjo and Okada (1974)
Silicoflagellates	$10^3/l$	Water, warm region	Travers and Travers (1968)
Euglena	$2 \times 10^2–2 \times 10^4/l$	Water, Norway	Throndsen (1969)
Yeasts	$10^6–10^8/l$	Neuston	Crow *et al.* (1975)
	10–8400/l	Water, Finland	Väätänen (1980)
	1–200/l	Water, Southern Ocean	Fell (1974)

Table 3.6. *Number of flagellates in seawater, as determined by direct counts using a haemocytometer*

Taxon	No. of organisms (/ml)
Heterotrophs	
Bicoeca sp.	2.15×10^3
Bodo spp.	11.28×10^3
Paraphysomonas sp.	2.72×10^3
Rhyncomonas sp.	2.91×10^3
Phototrophs	
Chroomonas salina	7.7×10^3
Dunaliella tertiolecta	6.12×10^3
Isochrysis galbana	4.09×10^3
Micromonas spp.	82.8×10^3
Monochrysis lutheri	3.74×10^3

Data from Caron (1983)

data on fungi is missing due to a lack of suitable methods of measurement (Kohlmeyer and Kohlmeyer, 1979). For that matter, all the predominant marine eukaryotic components of the picoplankton are imprecisely identified. According to Kohlmeyer (1983) the major factors controlling the distribution of fungi in the marine environment are the availability of suitable substrates (animate and inanimate), temperature, oxygen and pressure. Mycelial fungi occur on most surfaces, e.g. driftwood, exposed to seawater. The hyphae spread rapidly with the ultimate production of many spores. In the coastal environment, moulds are widespread in the vicinity of oil refineries (Grüttner and Jensen, 1983). Fungi are rare in the depths of the oceans (Ahearn and Crow, 1986). Yeasts are ubiquitous in the sea (Hagler and Ahearn, 1986), with up to $10^8/l$ present in neuston (Crow *et al.*, 1975). Offshore, increased densities of yeasts may be attributed to algal blooms (Meyers *et al.*, 1967).

Protozoa, which may be enumerated by epifluorescence microscopy (Sherr and Sherr, 1983), are planktonic, occurring in the open sea usually within 100 m of the water surface where abundant photosynthesising organisms are present. In deeper water, radiolaria are almost exclusively present (Sleigh, 1973). Amoebae are quite common in neuston, reaching populations of 1.2×10^3–1.35×10^3 cells/l (Sieburth *et al.*, 1976). On marine snow in the Sargasso Sea, Caron *et al.* (1982) estimated amoebae populations of up to $2.2 \times 10^4/l$; these numbers were much higher than in the surrounding water. Ciliates, which may comprise 4–57% of the total biomass of the heterotrophic nanoplankton (Sherr *et al.*, 1986), have been reported in densities of 1.4–162/ml in coastal waters (Andersen and Sørensen, 1986).

Large populations of flagellates may occur, even visible blooms. Principally, dinoflagellates are present in the upper layers of water, i.e. to depths of approximately 30 m, but not necessarily on the surface (Allen, 1941; Hasle, 1950). They occur in the open sea as well as coastal zones. North of the Antarctic Convergence Zone, dinoflagellates comprise the dominant component of the phytoplankton. Interestingly, south of the Antarctic Convergence, diatoms predominate (Hanson *et al.*, 1983). Indeed diatoms, which are restricted to the photic zone, may also form visible blooms (Sieburth, 1979).

4   Taxonomy of marine micro-organisms

A diverse range of micro-organisms, including bacteria, filamentous fungi, yeast, micro-algae (i.e. diatoms) and protozoa, is regarded as true inhabitants of the marine environment. Excluding transient organisms, and the specialised groups associated with the deep sea (Chapter 7) and/or pathogenesis of macro-organisms (Chapter 6), representatives of >100 prokaryotic genera may be described as marine inhabitants (Table 4.1). These organisms will be considered separately below.

Table 4.1. *Bacterial genera indigenous to the marine environment*

Category	Family	Genus
Phototrophs	Chromatiaceae	*Chromatium, Thiospirillum, Thiocystis*
	Thiocapsaceae	*Thiocapsa*
	Chlorobiaceae	*Chlorobium, Prosthecochloris*
	Related genus	*Chloroherpeton*
	Ectothiorhodospiraceae	*Ectothiorhodospira*
	Rhodospirillaceae	*Rhodocyclus, Rhodomicrobium, Rhodopseudomonas, Rhodospirillum*
	Related genus	*Erythrobacter*
Cyanobacteria	(Chroococcacean)	*Synechocystis, Synechococcus*
	(Nostocacean)	*Nostoc*
	(Oscillatorian)	*Lyngbya, Oscillatoria, Plectonema, Spirulina, Trichodesmium*
	(Pleurocapsalean)	*Dermocarpa*
	(Rivularian)	*Calothrix*
		Dichothrix, Richelia, Haliarachne, Dactyliococcopsis, Katagnymene, Nodularia, Pelagothrix
Other phototrophic prokaryotes	Prochloraceae	*Prochloron*
Gliding bacteria	Cytophagaceae	*Cytophaga, Flexibacter, Flexithrix, Herpetosiphon, Saprospira, Sporocytophaga*
	Related organisms	*Microscilla*
	Beggiatoaceae	*Beggiatoa, Thioploca*
	Leucotrichaceae	*Leucothrix, Thiothrix*
Budding and appendaged bacteria	Caulobacteraceae	*Caulobacter*
	Hyphomicrobiaceae	*Hyphomicrobium, Hyphomonas, Pedomicrobium*
	Planctomycetaceae	*Planctomyces, Pirella, Prosthecomicrobium*

Table 4.1 *(cont.)*

Category	Family	Genus
Aerobic/micro-aerophilic non-motile/ motile helical/vibrioid Gram-negative bacteria	Spirillaceae Spirosomaceae	*Bdellovibrio, Oceanospirillum* *Flectobacillus*
Gram-negative aerobic rods and cocci	Halobacteriaceae Methylococcaceae Neisseriaceae Pseudomonadaceae Alcaligenaceae —	*Halococcus, Halobacterium* *Methylophaga* *Acinetobacter* *Pseudomonas* *Alcaligenes,* *Alteromonas, Chromobacterium, Deleya, Flavobacterium, Janthinobacterium, Marinomonas, Paracoccus, Shewanella, Halomonas*
Facultatively anaerobic rods	Enterobacteriaceae Vibrionaceae	***Serratia*** ***Photobacterium, Vibrio, Listonella***
Gram-negative anaerobic rods and cocci	Haloanaerobiaceae Desulfurococcaceae	*Halobacteroides* *Desulfobacter, Desulfobulbus, Desulfococcus, Desulfosarcina, Desulfuromonas, Desulfovibrio*
Gram-negative chemolithotrophs (Ammonia- or nitrite-oxidising bacteria)	Nitrobacteraceae	*Nitrobacter, Nitrococcus, Nitrosococcus, Nitrospina, Nitrosomonas, Nitrosospira, Nitrospira*
(Sulphur bacteria)	— Achromatiaceae	*Macromonas, Thiobacillus, Thiomicrospira, Thiospira, Thiovulum, Thiobacterium* *Achromatium*
Methane bacteria	Methanobacteriaceae — Methanococcaceae Methanomicrobiaceae Methanoplanaceae Methanosarcinaceae	*Methanobacterium* *Methanospirillum* *Methanococcus* *Methanococcoides, Methanogenium, Methanomicrobium* *Methanoplanus* *Methanosarcina*
Gram-positive cocci	Micrococcaceae Planococcaceae	*Micrococcus, Staphylococcus* *Marinococcus*
Endospore-forming rods and cocci	Bacillaceae Clostridiaceae	*Bacillus* *Clostridium*
Actinomycetes and related bacteria	Actinomycetaceae Micromonosporaceae Nocardiaceae Mycobacteriaceae Streptomycetaceae —	*Actinomyces* *Micromonospora* *Nocardia, Rhodococcus* *Mycobacterium* *Streptomyces* Coryneforms (*Arthrobacter, Brevibacterium, Corynebacterium, Curtobacterium*)
Spirochaetes	Spirochaetaceae	*Crispispira, Spirochaeta*

4.1   Prokaryotes

The majority of the bacteria located in the water column comprise Gram-negative rods, with higher proportions of Gram-positive organisms occurring in the sediments. In terms of common traits, many of these organisms produce pigmented colonies (particularly yellow, orange or red) on solid media. Additionally, there is usually a requirement for NaCl, a pronounced exo-enzyme-producing potential, e.g. for DNA, and strains prefer utilising amino acids to sugars. Otherwise, phototrophs, autotrophs and heterotrophs appear to co-exist in comparatively close proximity. Their recovery from marine samples will reflect the nature of the methodology used (see Chapter 2).

It is notoriously difficult to identify fresh isolates from marine samples. This problem may suggest that, as yet, many novel groups remain to be properly characterised, and thus incorporated into diagnostic schemes. Alternatively, there is a notable difference in the characteristics of fresh compared to stored isolates. Conceivably, plasmid DNA, which is responsible for encoding many of the traits commonly used for diagnosis, may be lost on storage. Seemingly, fresh isolates have much greater enzymatic activity than their counterparts, which have been stored on laboratory media. Such differences in characteristics explain why named reference strains rarely cluster with environmental isolates in numerical taxonomy studies (e.g. Austin, 1982). Within these constraints, the taxonomy of marine prokaryotes will be discussed.

4.1.1   *Phototrophs containing bacteriochlorophyll*

Gram-negative phototrophs are common in surface waters. Yet unfortunately, the taxonomy of the organisms is best described as in a state of flux. Rightly or wrongly, great taxonomic weight has been placed on the nature of the photosynthetic pigment, e.g. bacteriochlorophyll *a* or *b*. The so-called purple photosynthetic bacteria, which carry out anoxygenic (does not form oxygen) photosynthesis, contain bacteriochlorophyll *a* or *b*. The green photosynthetic bacteria, which also conduct anoxygenic photosynthesis, contain only bacteriochlorophyll *a*. In contrast, the cyanobacteria (i.e. the so-called blue-green algae) and *Prochloron* contain chlorophyll *a*, and carry out oxygenic (oxygen-generating) photosynthesis. Dealing initially with the bacteriochlorophyll-containing bacteria, representatives of five families, i.e. Chlorobiaceae (green sulphur bacteria), Chromatiaceae (purple sulphur bacteria), Ectothiorhodospiraceae (purple sulphur bacteria), Rhodospirillaceae (purple non-sulphur bacteria) and Thiocapsaceae (purple sulphur bacteria) (Table 4.1) have been associated with the marine environment.

Of the family Chlorobiaceae, representatives of *Chlorobium* and *Prosthecochloris* may be found in marine situations. The former are non-motile anaerobic curved or straight rods without gas vacuoles, but which contain bacteriochlorophyll *c*, *d* or *e*, and the carotenoids, chlorobactene and isorenieratene. The pigments, giving rise to a yellow-green-brown cell mass, are contained in vesicles (Fig. 4.1) which lie beneath (and attached to) the cytoplasmic membrane. Photosynthesis occurs in the presence of hydrogen sulphide and during the process sulphur may be deposited as droplets externally to the cell. The guanine plus cytosine (G+C) ratio of the DNA is in the range of 49–58.1 moles % (Trüper and Pfennig, 1981), which indicates a level of heterogeneity in the species description. Marine isolates have been equated with *C. limicola* and *C. vibrioforme*. A second genus, *Prosthecochloris* (G+C ratio of the DNA = 50.0–56.0 moles %), differs from the above description insofar as cells are spherical, in chains and with prosthecae (individual cell dimensions are 0.3–0.7 μm × 0.5–1.2 μm). Bacteriochlorophyll *c* or *e* together with the carotenoids chlorobactene and

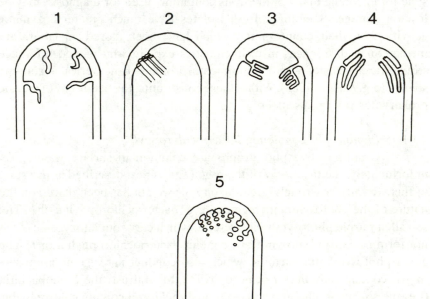

Fig. 4.1 Diagrammatic representation of the arrangement of intracytoplasmic membranes found in photosynthetic bacteria. 1 = tubes, which are found in *Rhodocyclus*, *Rhodopseudomonas* and *Rhodospirillum*; 2 = bundled tubes as present in *Thiocapsa*; 3 = stacks, which are located in *Ectothiorhodospira* and *Rhodospirillum*; 4 = membranes as found in *Rhodomicrobium* and some species of *Rhodopseudomonas*; 5 = vesicles, which are common to *Chromatium*, *Rhodopseudomonas*, *Rhodospirillum*, *Thiocapsa*, *Thiocystis* and *Thiospirillum* (after Trüper and Pfennig, 1981).

isorenieratene are contained in vesicles. Both species, i.e. *P. aestuarii* and *P. phaeoasteroidea*, occur in coastal and estuarine muds.

From the family Chromatiaceae, cells of *Chromatium, Thiocystis* and *Thiospirillum* occur in marine sites. With its many constituent species, *Chromatium* is extremely heterogeneous as illustrated by the spread of G+C values of the DNA from 48.8–70.4 moles %. Apparently, species comprise anaerobic unvacuolated slime-producing rods of 1–4.5 μm × ≤15 μm in size, which if motile possess polar flagella. Hydrogen sulphide is required for photosynthesis, whereupon sulphur droplets are stored intra-cellularly. Cell masses appear as various shades of purple or brown (Trüper and Pfennig, 1981). In contrast, *Thiocystis* comprise spherical cells of approximately 3.0 μm in diameter, containing okenone and/or rhodopinal as carotenoids, which gives a purple-violet-red coloration to the cell mass. The G+C ratio of the DNA has been calculated as 61.3–67.9 moles %. Both *T. gelatinosa* and *T. violacea* occur in coastal hydrogen sulphide-con-taining muds and waters. *Thiospirillum jenense* (G+C ratio of the DNA = 45.5 moles %) consists of spiral cells of 2.5–4.5 μm × 30–40 μm in size, which contain lycopene and rhodopin as carotenoids, and appear orange-brown *en masse*.

The genus *Ectothiorhodospira* (G+C ratio of the DNA = 52.9–68.4 moles %) comprises anaerobic spiral unvacuolated cells, which if motile possess polar flagella. Bacteriochlorophyll *a* or *b* is contained in stacked membranes (Fig. 4.1), and cell masses appear green or red. Hydrogen sulphide is oxidised during photosynthesis, during which sulphur droplets are deposited externally to the cell. *E. halochloris, E. halophila* and *E. mobilis* occur in coastal (enclosed) waters and on beaches (Trüper and Imhoff, 1981).

Of the family Rhodospirillaceae, recovery of representatives of *Rhodocyclus, Rhodomicrobium, Rhodopseudomonas* and *Rhodospirillum* is commonplace. *Rhodocyclus*, with *R. purpureus*, comprises intricately shaped micro-aerophilic, non-motile, purple-violet-pigmented cells, which may be described as circular or semi-circular in appearance. These cells are of 0.6–0.7 μm × 2.7–5.0 μm in size. The carotenoids include rhodopin and rhodopinal. Photosynthetic pigments are contained in intracytoplasmic membranes, arranged as tubes (Fig. 4.1). The G+C ratio of the DNA is 65.3 moles % (Trüper and Pfennig, 1981).

Rhodomicrobium contains one validly described species, i.e. *R. vannielii*, of Gram-negative anaerobic phototrophic prosthecate budding bacteria which are able to carry out oxidative metabolism under micro-aerophilic and aerobic conditions. The organisms possess a lamellar membrane system (Fig. 4.1), and contain bacteriochlorophyll *a*, group I carotenoids and βcarotene. The G+C ratio of the DNA is in the range of 61.8–63.8 moles % (Moore, 1981).

Rhodopseudomonas has two species which have been recovered from seawater, namely *R. marina* (Imhoff, 1983) and *R. sulfidophila* (Hansen and Veldkamp, 1973). These are regarded as purple non-sulphur bacteria, tolerating low concentrations of sulphide which is not oxidised to sulphate. Instead, thiosulphate and sulphur are the only products of oxidation. Cultures comprise short rods, which are motile by means of a flagellar pattern intermediate between lateral, polar and peritrichous. Photosynthetic pigments, i.e. bacteriochlorophyll *a* and carotenoids of the spirilloxanthine series, are contained in intracytoplasmic membranes, arranged as stacks (Fig. 4.1), and lying parallel to the cytoplasmic membrane.

Rhodospirillum, represented by the obligatory halophilic species *R. salexigens*, comprises curved or spiral-shaped Gram-negative cells which are motile by means of bipolar flagella. Intracytoplasmic membranes, arranged parallel to the cytoplasmic membrane, contain bacteriochlorophyll *a* and spirilloxanthine as the major pigments. The G + C ratio of the DNA is 64 moles % (Drews, 1981).

Thiocapsa (G+C ratio of the DNA = 63.3–69.9 moles %), with *T. pfennigii* and *T. roseopersicina*, is found in estuarine and coastal muds. Cultures comprise orange-brown-pink-red-pigmented non-motile unvacuolated spherical cells of 1.2–3.0 μm in diameter, which form aggregates in slime. The carotenoids are spirilloxanthine and tetrahydrospirilloxanthine. Together with bacteriochlorophyll *a* and *b*, these are contained in intracytoplasmic membranes of vesicular or tube shape (Fig. 4.1).

Two other genera, namely *Chloroherpeton* and *Erythrobacter*, remain to be allocated to family groupings. *Chloroherpeton*, with its only species, *C. thalassium*, is a long rod-shaped Gram-negative, gliding, green sulphur organism. It is an obligate phototroph with bacteriochlorophyll *c* and a small quantity of bacteriochlorophyll *a* together with γ-carotene, requiring CO_2 and sulphide for growth. Sulphur is deposited outside of the cells. Nitrogen may be fixed. The G + C content of the DNA is 45.0–48.2 moles % (Gibson *et al.*, 1984). Although thought to be related to Rhodosprillaceae, the monospecific genus *Erythrobacter*, with *E. longus* as the validly described species, does not grow phototrophically. However, the cells contain bacteriochlorophyll *a*. The oval rods are motile by sub-polar flagella, aerobic, require biotin, produce catalase, oxidase and phosphatase, degrade gelatin and Tween 80, and utilise glucose, acetate, butyrate, glutamate and pyruvate as the sole sources of carbon. The G + C ratio of the DNA is 60–64 moles % (Shiba and Simidu, 1982).

4.1.2   *Cyanobacteria*

Unlike other bacteria, the taxonomy of the cyanobacteria (alias blue-green algae/bacteria) is controlled under the terms of the Botanical Code. Consequently, it would appear that morphology, used extensively in plant systematics, has exerted a profound influence on the taxonomy of the cyanobacteria. Taxonomically, the cyanobacteria comprise a diverse range of unicellular and filamentous prokaryotes, which carry out oxygenic photosynthesis using water as an electron donor, with the production of gaseous oxygen in light. Chlorophyll *a* is contained in paired photosynthetic lamellae, termed thylakoids, the outer surfaces of which possess granules (phycobilisomes) comprised of phycobiliprotein pigments. Motility, if exhibited, is by gliding and never by flagella. Intracellular gas vacuoles, which function for buoyancy, may be present. These are formed of gas vesicles, which are surrounded by a proteinaceous membrane. Unicellular cyanobacteria divide by binary or multiple fission, or by release of apical cells, termed exospores. Filamentous forms multiply by fragmentation or by release of short, motile, possibly sheathed, loosely-attached chains of cells, i.e. trichomes/hormogonia. Some filamentous cyanobacteria produce specialised cells, called akinetes and heterocysts. The former, being larger than the vegetative cells, constitute a thick-walled resting stage. Akinetes eventually germinate to produce short trichomes. In contrast, heterocysts do not function in reproduction. These cells contain refractile polar granules and a thick cell wall. It has been concluded that in aerobic conditions, heterocysts serve as sites for nitrogen fixation (Stanier, 1974; Rippka *et al.*, 1981; Walsby, 1981; Waterbury and Stanier, 1981).

Chroococcacean cyanobacteria are the simplest morphologically, comprising mostly non-motile unicellular rods and cocci, which multiply by binary fission or budding. Moreover, many taxa have now been isolated in pure culture. *Synechococcus*, which is rod-shaped, possesses thylakoids and has a G+C ratio of the DNA of 39–71 moles % (this is obviously very heterogeneous), occurs commonly in the ocean where population densities may reach as high as 3×10^5 cells/ml (Waterbury and Stanier, 1981). A second example, found in the marine environment is the genus *Synechocystis* (G+C ratio of the DNA = 35–48 moles %), which comprises coccoid cells devoid of a sheath (Rippka *et al.*, 1981).

Pleurocapsalean cyanobacteria are also unicellular, but form vegetative aggregates. Reproduction is by small spherical baeocytes (endospores), which glide. Most abundant in temperate and tropical regions, pleurocapsalean cyanobacteria are common epiphytes.

The open ocean possesses a narrow range of planktonic forms, typified by the genus *Trichodesmium*, which forms dense blooms, particularly in the tropics and sub-tropics. *Trichodesmium*, with its filaments but without heterocysts, probably fixes nitrogen, although this is difficult to confirm because the organism has not been successfully cultured (Waterbury and Stanier, 1981). *Trichodesmium*, notably *T. contortum, T. erythraeum* and *T. thiebautii*, comprises the only genus with marine forms to possess gas vacuoles. These are arranged as peripheral cylinders, and do not collapse below 10 bars of pressure. The genus is considered to be related to *Oscillatoria* (Walsby, 1981).

Oscillatorian cyanobacteria possess trichomes which are composed exclusively of vegetative cells. Intercalary cell division occurs at right angles to the long axis of the trichome. The genus *Oscillatoria* comprises filamentous sheathed cells. The trichome of the disc-shaped cells is always motile. The G+C ratio of the DNA appears to be in the range of 40–50 moles % (Walsby, 1981). Another representative is *Lyngbya*, i.e. *L. majuscula*, which also forms blooms in the sea. This taxon is considered to be closely related to *Plectonema*.

Nostocacean cyanobacteria, of which *Nostoc* is bloom-forming, are distinguished by the presence of a trichome of even width, and by heterocysts which often develop in terminal or intercalary positions. In contrast with rivularian cyanobacteria, e.g. *Calothrix*, the trichome tapers from base to apex. *Nostoc* has a G+C ratio of the DNA of 39–46 moles % (Rippka *et al.*, 1981).

Other cyanobacterial bloom-formers include *Katagnymene* spp. and *Nodularia spumigena* (Walsby, 1981). The latter comprises disc-shaped filamentous cells, with heterocysts occurring singly. The G+C ratio of the DNA is 41 moles %.

4.1.3   *Prochloron*

A novel group of photosynthetic prokaryotes containing chlorophyll *a* and *b* in paired thylakoids but lacking phycobilins, was discovered by Lewin (1976, 1977) and placed in a newly created algal division, termed *Prochlorophyta*, based on the genus *Prochloron*, with *P. didemni* as the sole species. The organism comprises unvacuolated uni- or multi-cellular Gram-negative aerobic or facultatively anaerobic cells, which perform oxygenic photosynthesis. The organism is motile and spherical, approximately 6–25 μm in diameter. Peptidoglycan is present in the wall. The G + C ratio of the DNA is in the range of 39–41 moles %. After its initial association with the algae, *Prochloron* was subsequently considered to be included with the photobacteria (Florenzano *et al.*, 1986), possibly evolved from typical cyanobacteria. The organism was originally recovered as a symbiont of colonial ascidians (Didemnidae). However, *Prochloron* may be found free-living in shallow seawater in tropical or sub-tropical conditions, i.e. ≥ 20 °C.

4.1.4 Gliding bacteria

The Gram-negative pigmented bacteria, which move by gliding across solid/liquid interfaces, encompass four family groupings. In particular, carotenoid pigment-containing cytophagas, which do not produce microcysts, are abundant in the coastal marine environment, where the bacteria will be found in seawater, on dead and living seaweed, on the surface of fish, and in sediments. Although the taxonomy of *Cytophaga* is highly confused and the speciation of marine isolates is unclear, it is apparent that there is a marked ability to degrade complex molecules, such as agar, cellulose and chitin. The organisms are strictly aerobic, mesophilic and grow readily on marine 2216E agar, although swarming, an indication of gliding motility, may not occur. For this phenomenon to be displayed, low-nutrient-containing media are advocated. The G+C ratio of the DNA lies in the range of 28–39 moles %. Confusion is likely to occur between *Cytophaga* and *Flexibacter*, the latter of which forms long flexible cells, has less degradative activity, and demonstrates G+C values of the DNA of 46–48 moles %. It has been argued, however, that *Flexibacter* should be restricted to soil and freshwater isolates, with *Microscilla*, notably *M. marina*, receiving the marine counterparts (Reichenbach and Dworkin, 1981). A morphologically related strictly aerobic yellow-pigmented organism has been classified in *Flexithrix*, as *F. dorotheae* (G+C ratio of the DNA = 37.5 moles %). This organism forms very long (~500 μm) filaments with false branching. Complex molecules, e.g. agar, cellulose, gelatin and starch, are not degraded. Nevertheless, the taxonomic status of this organism is unclear. Long (300–>1200 μm), stiff multicellular filaments with transparent 'sleeves' at the ends have been classified as *Herpetosiphon*. Yellow-orange colonies comprise fairly unreactive strictly aerobic cells. Marine strains, largely obtained from sediments, have been identified as *H. gigantus*, *H. cohaerens*, *H. persicus* and *H. nigricians*. *Saprospira*, the genus of which has not been studied extensively, comprises pink/yellow-orange helical, filamentous cells of 6–500 μm in size, with a G+C ratio of the DNA in the range of 38–48 moles %. Again, the organisms are strictly aerobic, and do not demonstrate marked degradative activity. *S. grandis* and *S. toviformis* inhabit sand and muds. A microcyst-forming marine organism belongs in the genus *Sporocytophaga* (G+C ratio of the

DNA = 36 moles %). Such strictly aerobic catalase-positive cells degrade cellulose. The only species studied in any detail is *S. myxococcoides* (Reichenbach and Dworkin, 1981).

Only a few strains of *Beggiatoa* (G+C ratio of the DNA = ~37 moles %) have been studied in axenic culture (Wiessner, 1981). Unfortunately, marine isolates seem difficult to speciate. These aerobic to micro-aerophilic chemoautotrophic filamentous organisms develop trichomes with slimy sheaths. When grown in the presence of hydrogen sulphide, sulphur droplets are deposited intracellularly. *Beggiatoa* is found in the vicinity of decaying seaweed. Some interest has focused on the genus *Thioploca*, of which *T. araucea*, *T. chileae* and *T. ingrica* have been associated with marine sediments (Maier, 1974; Maer and Gallardo, 1984). Briefly, the taxa comprise multicellular sheathed filamentous Gram-negative sulphur-containing cells, which have never been isolated in pure culture. Therefore, some doubt must surely be cast upon the taxonomic meaning of the group.

The monospecific genus *Leucothrix* (with *L. mucor*; G+C ratio of the DNA = 47–49 moles %) comprises very large cells of 2–3 μm in diameter and 1000–1500 μm in length, which undergo a life cycle. The strictly aerobic filaments round up to become ovoid or spherical gonidia that demonstrate a type of gliding movement when released. These gonidia may aggregate to form rosettes. *L. mucor* may be recovered as an epiphyte on marine algae, or as an epibiont on shellfish. The bacterial taxon may be closely related to the sulphur droplet-containing *Thiothrix*, which has also been linked to the cyanobacteria (Brock, 1981). *Thiothrix* (*T. nivea*) occurs in sulphide-rich areas, such as in the vicinity of rotting seaweed.

4.1.5   *Budding and appendaged bacteria*

The morphologically intriguing Gram-negative bacteria classified in the families Caulobacteraceae, Hyphomicrobiaceae and Planctomycetaceae, abound, particularly in the coastal marine environment. Colonies may be regularly isolated using common media e.g. marine 2216E agar (Difco) and aerobic incubation at 15–25 °C. Two species of the appendaged 'rosette-forming' *Caulobacter*, i.e. *C. halobacteroides* and *C. maris*, occur in seawater (Poindexter, 1964). *Hyphomicrobium* and *Hyphomonas* may be primary colonisers of barren surfaces. *Hyphomonas* comprises aerobic prosthecate budding bacteria, which are motile by means of a single polar flagellum. The cell size and shape varies considerably during the growth cycle. Indeed, it may be argued that the organism undergoes a primitive life cycle. Cells contain polyphosphate and polyhydroxybutyrate, produce catalase and oxidase but not arginine dihydrolase, indole, or lysine or ornithine decarboxylase, and do not attack DNA, gelatin, starch or urea. Nutritionally, there is a definite preference for amino acids rather than sugars for growth. Species associated with surface waters include *H. neptunium* (this was formerly *Hyphomicrobium neptunium*), *H. oceanitis* and *H. polymorpha* (Weiner *et al.*, 1985). Additional species are present in the deep sea. A closely related genus is *Hyphomicrobium* (*H. vulgare* is present in marine waters), the cell shape and size of which does not alter during the growth cycle (Moore, 1981).

The two genera, *Pirella* and *Planctomyces*, classified in the family Planctomycetaceae (Schlesner and Stackebrandt, 1986), contain species which are native to the sea. *Pirella marina* is a salt-requiring fimbriated budding, non-prosthecated, rosette-forming organism, which is egg-shaped or elliptical, and possesses crateriform structures. Motility is by means of a single polar flagellum. The G + C ratio of the DNA is in the range of 53.6–57.4 moles % (Schlesner, 1986). A similar description is befitting of *Planctomyces maris*, which is budding and stalked, and motile by a single polar flagellum. Here, the G + C ratio is somewhat lower, being recorded as 50.5 moles % (Bauld and Staley, 1976; 1980).

The multiple-appendaged prosthecate, non-motile, unpigmented aerobic *Prosthecomicrobium litoralum* is of coastal origin, particularly in the vicinity of plants, e.g. *Ulva* (Bauld *et al.*, 1983). The G + C ratio of the DNA has been calculated as 66–67 moles %.

4.1.6   *Aerobic Gram-negative rods and cocci*

These organisms are among the most common groups recovered from marine samples on conventional nutrient-rich media. Two genera, i.e. *Bdellovibrio* and *Oceanospirillum*, are representative of the family Spirillaceae. *Bdellovibrio* is particularly fascinating insofar as strains are parasitic on other Gram-negative cells. The small comma-shaped cells of 0.2–0.5 μm $\times$ 0.5–1.4 μm in size are vigorously motile by means of a single polar flagellum. The metabolism is strictly respiratory, and along with many other marine organisms, catalase, gelatinase and oxidase are produced, and nitrate is reduced. The G + C ratio of the DNA lies in the range of 33.4–38.6 moles % (Burnham and Conti, 1984). It seems likely that marine isolates belong to as yet undescribed species. Guelin *et al.* (1977) reported the existence of other bdellovibrio-like organisms, i.e. micro-vibrios. These were coined *Microvibrio marinus roscoffensis*. However, the name has not stuck in the marine literature.

The genus *Oceanospirillum* was established out of a sub-division of *Spirillum* (Krieg, 1984). Many marine species have been described, including *O. beijerinckii, O. communa, O. hiroshimense, O. jannaschii, O. japonicum, O. kriegii, O. linum, O. maris* subsp. *maris, O. maris* subsp.

williamsae, O. minutulum, O. multiglobuliferum, O. pelagicum, O. pusil-lum and *O. vagum*. Typically, growth occurs aerobically in two to three days on simple defined media with amino acids or the salts of organic acids as the carbon source. Colonies comprise Gram-negative helical cells which are motile by bipolar flagella occurring singly or in tufts. Poly-β-hydro-xybutyrate is stored intracellularly. Coccoid bodies occur in old cultures. Oxidase but not indole is produced, and the catalase test generates a variable result. Starch and casein are not attacked, and sugars are not catabolised. The metabolism is respiratory. The G + C ratio of the DNA is in the range of 42–51 moles %. Isolates are common (comprising between 0.1 and 2.5% of the total bacterial microflora) in coastal seawater and decaying seaweed and animals. It is unclear whether or not oceanospirilla occur in the open ocean.

Representing the family Spirosomaceae, the genus *Flectobacillus* com-prises isolates which are found in seawater. Thus, *F. marinus* comprises ringed, comma-shaped or straight Gram-negative non-motile obligately aerobic rods, which form colonies with a pinky hue. Both catalase and oxidase but not indole or urease are produced. Nitrates are not reduced. Casein, cellulose, chitin, gelatin and starch are not attacked. Nevertheless, many compounds are utilised as the source of carbon, including acetate and citrate. The G + C ratio of the DNA is in the range of 34–38 moles % (Larkin and Borrall, 1984). Interestingly the taxon was formerly classified in the genus *Microcyclus*.

Halophilic organisms, i.e. needing 15% (w/v) sodium chloride, are associated with the family Halobacteriaceae, of which representatives of *Halobacterium* and *Halococcus* occur in the marine environment. *Halobac-terium* (marine isolates include *H. denitrificans* , *H. mediterranei, H. phar-conis, H. saccharovorum, H. salinarium, H. sodomense* and *H. volcanii*), comprises Gram-negative motile or non-motile rods, which possess a respiratory metabolism, and produce both catalase and oxidase. Pink, red or orange colonies are produced. The best growth occurs in 20–26% (w/v) sodium chloride. The G+C ratio of the DNA has been reported as 63–68 moles % (Larsen, 1984). Similar traits have been reported for *Halococcus* (marine isolates are classified as *H. morrhuae*). This is a pink, red or orange-pigmented organism, which comprises non-motile Gram-negative catalase- and oxidase-producing cocci. Cell division is by septation. The metabolism is respiratory, and nitrates are reduced. A slightly lower G+C ratio of 61–66 moles % has been recorded (Larsen, 1984).

Marine isolates of the genus *Methylophaga* (G + C ratio of the DNA = 38–46 moles %), include *M. marina* and *M. thalassica*. These are slightly aerobic, methylotrophic small Gram-negative rods of 1.0 μm × 0.2 μm in size, which are motile by a single polar flagellum. There is a requirement for vitamin B_{12}. Methylamine is utilised as both a carbon and nitrogen source. Seemingly, the organisms may be recovered from algae and sediment (Janvier *et al.*, 1985).

Fairly unreactive, difficult to identify Gram-negative non-motile catalase-positive but oxidase-negative rods have been lumped in the genus *Acinetobacter*, the species composition of which is unclear (Juni, 1984; Bouvet and Grimont, 1986), and the closely related *Moraxella* (Nishimura *et al.*, 1986).

It is debatable whether or not representatives of the genus *Pseudomonas* constitute *bona fide* marine organisms. However, it is clear that salt-requiring nutritionally diverse pseudomonads exist, which possess a respiratory metabolism, produce both catalase and oxidase, and are motile by polarly located flagella. True marine species include *P. doudoroffi*, *P. elongata*, *P. gelidicola*, *P. nautica*, *P. perfectomarina* and *P. stanieri* (Baumann *et al.*, 1983). In addition, isolates of *P. fluorescens* may also abound in coastal waters, although it seems likely that they are of terrestrial or freshwater origin.

Many genera have not been included in family groupings. Nevertheless, some of these genera contribute significantly to marine prokaryotic biomass. The first example is the genus *Alteromonas*, which is in a state of taxonomic flux insofar as two species, namely *A. hanedai* and *A. putrefaciens*, have been recommended for inclusion in the newly described genus, *Shewanella* (MacDonell and Colwell, 1985). Moreover, from rRNA studies, Van Landschoot and De Ley (1983) noted four separate branches among *Alteromonas* species. One of these branches, which comprised *A. communis* and *A. vaga*, was reclassified in a newly created genus, *Marinomonas*, as *M. communis* and *M. vaga*, respectively. Marked variation is apparent among the remaining species of *Alteromonas*, particularly in colonial pigmentation of which red (*A. rubra*), orange (*A. aurantia*), lemon-yellow (*A. citrea*) and violet-coloured (*A. luteoviolacea*) taxa has been described. Other marine species include *A. espejiana*, *A. haloplanktis*, *A. macleodii* and *A. undina*. Briefly, alteromonads (G + C ratio of the DNA = 38–50 moles %) comprise Gram-negative rods which are motile by a single polar flagellum. The metabolism is respiratory, and a diverse range of metabolic activity is apparent (Baumann *et al.*, 1984a). The organisms are common in coastal water and the open ocean.

Purple pigmented, strictly aerobic Gram-negative rods which are motile by means of a single polar flagellum may also belong to the genera

Chromobacterium and *Janthinobacterium*. Certainly, *J. lividum* occurs in low numbers in coastal waters. However, the taxonomic status of *C. marinum* is confused (Hamilton and Austin, 1967).

The genus *Deleya* comprises marine isolates, formerly classified as *Alcaligenes* (Kersters and De Ley, 1984). These are strictly aerobic Gram-negative rods (the G + C value of the DNA is 52–68 moles %), which are motile by means of a degenerately peritrichous arrangement of flagella. The marine species include *D. aesta, D. cupida, D. marina* (this organism was previously classified in the genus *Pseudomonas* as *P. marina*), *D. pacifica* and *D. venusta*.

Another group which may be considered to be a taxonomist's nightmare is *Flavobacterium*. This comprises non-motile, pigmented Gram-negative rods, with a G + C ratio of the DNA in the range of 31–42 moles % (Holmes *et al.*, 1984). Marine species have been described, but these are of dubious validity. Thus, *F. halmophilum* and *F. oceanosedimentum* have G + C values of 49.7 and 67.5 moles %, respectively, *F. okeanokoietes* is motile by peritrichous flagella, *F. marinotypicum* has a Gram-positive staining reaction, and *F. uliginosum* is probably a cytophaga.

Paracoccus and *Halomonas* (G + C ratio of the DNA = 60.5 moles %) may possibly be resident in the marine environment. The former comprises Gram-negative spherical or short rod-shaped cells, which are non-motile and produce both catalase and oxidase (Kocur, 1984). The latter (i.e. *H. elongata*) consists of Gram-negative rods, which are motile by polar flagella. Sodium chloride at 0.1–32.5% (w/v) is tolerated (Vreeland, 1984).

4.1.7   *Facultatively anaerobic Gram-negative rods*

Despite the comparative abundance of Enterobacteriaceae representatives in freshwater and terrestrial habitats, only one species may be considered as a marine inhabitant, albeit rarely encountered. Originally, the organism was classified as *Serratia marinorubra* (Zobell and Upham, 1944) although it is now known as *Serratia rubidaea*. Essentially, it is a red-pigmented, Gram-negative rod-shaped organism, which produces catalase but not oxidase, and is motile by peritrichous flagella (Grimont and Grimont, 1984).

The marine environment is home for a fascinating collection of light-emitting organisms, of which representatives of the genus *Photobacterium*, notably *P. angustum, P. leiognathi* and *P. phosphoreum*, are prime examples. These sodium-requiring organisms are common in seawater and on marine animals. For example, photobacteria appear as virtually pure cultures on the light-emitting organs of some fish. Briefly, they are fermentative, and motile by polar flagella (Baumann and Baumann, 1984).

Among the best studied inhabitants of the coastal and estuarine environments, the genus *Vibrio* is almost synonymous with the definition of a true marine organism. However, judgement on this pronouncement will be reserved. Nevertheless, new species are regularly named. Currently, 'marine' species include *V. aestuarianus, V. alginolyticus, V. campbellii, V. damsela* (fish pathogen), *V. diazotrophicus* (nitrogen-fixing organism), *V. fischeri* (yellow-orange-pigmented light-emitting organism), *V. gazogenes* (red-pigmented colonies), *V. harveyi* (formerly classified as *Lucibacterium harveyi*), *V. logei* (yellow-orange-pigmented colonies), *V. marinus, V. mediterranei, V. natriegens, V. nereis, V. nigripulchritudo* (blue-black-pigmented colonies), *V. orientalis, V. parahaemolyticus, V. splendidus* and *V. vulnificus*. After detailed RNA cataloguing MacDonell and Colwell (1985) transferred *V. pelagius* into the newly described genus *Listonella*, as *L. pelagius*. This may be seen as a wise move since the G + C spread of *Vibrio* of 38–51 moles % was not conducive to the notion of a homogeneous genus. Historically, there has been heated debate over the validity of *Vibrio* and *Beneckea*, the latter of which was used for over a decade to describe the same species. Thus, the literature abounds with references to the same specific epiphet appearing under the guise of *Vibrio* or *Beneckea*. Fortunately, to eliminate further confusion *Beneckea* was formally withdrawn, and the species epiphets transferred/reverted to *Vibrio*. So *Vibrio* may be considered as consisting of Gram-negative, fermentative, catalase- and oxidase-producing rods, which are motile by polar flagella. In addition, most strains are sensitive to the vibriostatic agent, O/129 (Baumann *et al.*, 1984b).

4.1.8 *Gram-negative anaerobic rods and cocci*

Oren *et al.* (1984) described a new genus of moderately halophilic anaerobic bacteria, recovered from the sediment in the Dead Sea. The organisms were classified as a monospecific genus, *Halobacteroides* (*H. halobius*), in the family Haloanaerobiaceae. Essentially, the organisms comprise fairly unreactive long Gram-negative rods, which are motile by peritrichous flagella. Neither catalase nor oxidase is produced. The G + C ratio of the DNA is 30.7 moles %.

Strictly anaerobic, Gram-negative, sulphate-reducing organisms found in marine sediments have been classified in six genera of the family Desulfurococcaceae (Table 4.1). *Desulfobacter*, represented by *D. postgatei*, comprises ellipsoidal or rod-shaped cells, which are motile by a single polar flagellum (Widdell and Pfennig, 1984a). A similar description occurs for *Desulfobulbus*, of which *D. propionicus* occurs in marine muds (Widdell and Pfennig, 1984b). A slightly different description exists for *Desulfococ-*

cus, which is spherical and usually non-motile. The marine form is *D. multivorans* (Widdell and Pfennig, 1984c). *Desulfosarcina variabilis* comprises irregularly-shaped, usually non-motile cells (Widdell and Pfennig, 1984d). *Desulfovibrio* comprises rods, which are motile by polar flagella. Marine species include *D. africanus, D. baarsii, D. desulfuricans, D. gigas, D. salexigens* and *D. vulgaris* (Postgate, 1984). *Desulfuromonas*, with its marine species *D. acetoxidans*, forms peach-pink-pigmented colonies containing rods, which are motile by a single polar flagellum. The organism grows well in syntrophic cultures containing phototrophic green sulphur bacteria. However with a G + C range of 50–63 moles %, it would appear that the genus is heterogeneous (Pfennig, 1984).

4.1.9   *Gram-negative chemolithotrophs (ammonia- or nitrite-oxidising bacteria)*

These organisms have been grouped together according to their ability to oxidise inorganic compounds as a source of energy. Arguably, this parameter is too artificial for sound taxonomic practice. Generally, the bacteria have been categorised by size, shape and arrangement of the intracytoplasmic membranes. Four genera, namely *Nitrobacter, Nitrococcus, Nitrospina* and *Nitrospira*, contain nitrite-oxidising bacteria, of which the former is the most important in this role. Generally, these organisms occupy strictly aerobic habitats. Of these bacteria, *Nitrobacter* is a monospecific genus (G + C ratio of the DNA = 60.7–61.7 moles %), i.e. with *N. winogradskyi* as the sole member, comprising rods which reproduce by budding. If motile, isolates possess a single polar flagellum. There is a polar cap of cytomembranes. Of interest to microbiologists, the organism has an unique ultrastructure to the outermost layer of the cell envelope. Essentially, this comprises two uneven electron-dense layers. *Nitrococcus*, with *N. mobilis* as its sole species (G + C ratio of the DNA = 61.2 moles %) comprises spherical cells of $\geqslant 1.5$ μm in diameter. Motility is achieved by one or two polarly located flagella. With this taxon there is a tubular arrangement of the cytomembranes. In contrast, *Nitrospina gracilis* comprises long, slender non-motile rods which lack any extensive cytomembrane system. The G + C ratio of the DNA has been determined as 57.7 moles % (Watson *et al.*, 1981). The newly recognised member of the group is the monospecific genus *Nitrospira*, with *N. marina* (G + C ratio of the DNA = 50 moles %) as its sole species (Watson *et al.*, 1986). Although only two isolates have been studied in any detail, it has been concluded that cultures comprise strictly aerobic, yellow-brown-coloured, non-motile curved rods of 0.3–0.4 μm × 1.5 – 1.75 μm in size which form spirals with one to 12 turns. Division is by transverse fission. Similarly to *Nitrospina*, there is a distinctive multilayered cell envelope and a lack of cytomembranes. Despite the apparent brevity of the data, it was concluded that this organism possibly comprises one of the most common nitrite-oxidisers in the marine environment.

Three genera of ammonia-oxidising bacteria occur in the marine environment. *Nitrosococcus* (*N. mobilis* and *N. oceanus* occur in the sea) comprises large spherical cells of >1.5 μm in diameter, which are motile by means of one or more sub-polar flagella. Strains of *N. oceanus* possess an extra layer to the cell envelope. From the available information, it would appear that the G + C ratio of the DNA is 50.5–51.0 moles %. The genus *Nitrosomonas* has only one *bona fide* species, namely *N. europaea*. However, heterogeneity is apparent insofar as marine strains are clearly distinct to those of freshwater origin. For example, marine isolates possess an additional layer to the cell envelope, and demonstrate a requirement for salt. Nevertheless within such constraints, the taxon comprises short Gram-negative rods which may be motile by one or two sub-polarly positioned flagella. The G + C ratio of the DNA appears to be 47.4–51.0 moles %. *Nitrosospira*, with *N. briensis* (G + C ratio of the DNA = 54.1 moles %) as its sole species, is more often associated with soil than the aquatic environment. The cells are spiral-shaped, lack cytomembranes, and may be motile by degenerately peritrichous flagella (Watson *et al.*, 1981).

4.1.10   *Gram-negative chemolithotrophs (sulphur bacteria)*

Little is known about the taxonomy of the sulphur bacteria. *Macromonas*, which is often found in association with *Achromatium*, has never been cultured, and all observations on the organism have been made on natural material. The two species found in the marine environment are *M. bipunctata* and *M. mobilis*. Apparently, the colourless cells are cylindrical or bean-shaped of 9–20 μm in size, and are motile by a single polar flagellum. Daughter cells may be formed. Sulphur droplets and large spherical bodies, possibly of $CaCO_3$, occur intracellularly (La Rivière and Schmidt, 1981). Similarly, little is known about the biology of *Thiobacterium*. It would appear that *T. bovista* occurs in marine systems. This organism comprises non-motile rods of 2–5 μm × 0.6–1.5 μm in size with sulphur inclusions. The cells are inevitably embedded in gelatinous masses. *Achromatium* comprises spherical, ovoid or cylindrical cells with a length of between 5 and 100 μm. These Gram-negative catalase-negative obligate aerobes show slow jerking movement by peritrichous filaments. Most observations have been made on natural material, insofar as effective enrichment culture techniques have not been developed. Of the species associated with marine muds, *A. oxaliferum* contains massive inclusions of calcium carbonate together with some intracellular sulphur droplets, whereas *A. volutans* possesses many sulphur droplets but no calcium carbonate inclusions (La Rivière and Schmidt, 1981).

The genus *Thiobacillus* is evidently heterogeneous, from the wide range of G + C values reported for the species. Of the taxa found in the marine environment, i.e. *T. denitrificans, T. neapolitanus* and *T. thioparus*, the G + C values of the DNA have been reported as 63.0–67.9, 55.1–57.0 and 62.0–66.0 moles %, respectively. Effectively, these obligate chemolitho-trophs oxidise inorganic sulphur compounds, and assimilate carbon dioxide (Kuenen and Tuovinen, 1981). *Thiomicrospira* (*T. denitrificans* and *T. pelophila* occur in marine sediments) are very thin comma- or spiral-shaped organisms, which are motile by polar flagella. The G + C ratio of the DNA is approximately 48 moles % (Kuenen and Veldkamp, 1972). *Thiospira* spp. are spiral organisms, which may reach a length of 14 μm. Motility, if present, is by polar flagella. The two 'marine' species, i.e. *T. bipunctata* and *T. winogradskyi*, contain sulphur droplets, and exhibit chemotaxis to approp-riate oxygen–hydrogen sulphide gradients (La Rivière andSchmidt, 1981). Finally, a little more is known about the biology of *Thiovulum*, insofar as enrichment culture techniques have been developed successfully for *T. majus*. These ovoid, sulphur droplet-containing organisms of 5–25 μm in size divide by longitudinal fission which is preceded by constriction. Motility is by means of peritrichously positioned flagella. These catalase-negative, slime-excreting organisms possess thin cell envelopes and distinctive polar fimbrillar organelles. The bacteria are found in the interfaces of oxygen- and hydrogen sulphide-containing regions. Essentially, isolates are killed in highly oxygenated or totally anaerobic conditions (La Rivière and Schmidt, 1981).

4.1.11   *The methane bacteria*

There has been an upsurge of interest in the methane bacteria, with the consequence that new taxa are regularly described. Moreover, the descriptions are usually sound and lucid. Currently, marine representatives have been classified in five families, namely Methanobacteriaceae, Methanococcaceae, Methanomicrobiaceae, Methanoplanaceae and Methanosarcinaceae. *Methanobacterium* (G+C ratio of the DNA = 33–50 moles %) contains strictly anaerobic filamentous Gram-positive, non-motile fimbriated rods. The two species found in marine sediments include *M. formicicum* and *M. soehngenii*, the latter of which has never been isolated in pure culture (Balch *et al.*, 1979). *Methanospirillum*, with *M. hungatei* found in marine sediments, consists of small Gram-negative motile curved rods, which often form long spiral filaments. The G + C ratio of the DNA is 45–46.5 moles %. Five species of *Methanococcus* (G + C ratio of the DNA = 30.7–33 moles %), namely *M. jannaschi, M. maripaludis, M. thermolithotrophicus, M. vannielii* and *M. voltae*, have been found in marine sediments. These Gram-negative anaerobic cocci obtain energy for growth by oxidation of formate with the concomitant reduction of carbon dioxide to methane. The cells are motile by presumably polar flagella. Growth occurs optimally at 32–40 °C, and there is a requirement for 1.2–4.8% (w/v) sodium chloride (Balch *et al.*, 1979; Jones *et al.*, 1983). A morphologically related anaerobic organism, albeit in the family Methanomicrobiaceae, is *Methanococcoides methyluteus* (G + C ratio of the DNA = 42 moles %), which is also found in anaerobic marine sediments. This methanogen comprises non-motile irregular cocci of one μm in diameter (Sowers and Ferry, 1983). *Methanogenium*, notably *M. cariaci, M. marisnigri* and *M. thermophilicum*, are also characterised by the presence of irregular cocci 1.0–1.3 μm in diameter, which occur singly or in pairs. This Gram-negative non-motile methanogen apparently has a G + C ratio of the DNA of 59 moles % (Romesser *et al.*, 1979; Rivard and Smith, 1982). The type genus of the family is *Methanomicrobium* of which *M. mobile* has been recovered from marine sediments. This taxon comprises strictly anaerobic Gram-negative rods and cocci, which are motile by presumably polarly located flagella. The G + C ratio of the DNA is 48.8 moles % (Balch *et al.*, 1979). A morphologically fascinating methanogen belonging to the family Methanoplanaceae has been classified in the genus *Methanoplanus* as *M. limicola* (G + C ratio of the DNA = 47.5 moles %). The Gram-negative organism may be described as plate-shaped, being motile by polar tufts of flagella. It is notable that cells contain electron-dense inclusions. So far, it would appear that the normal habitat is most likely seawater (Wildgruber *et al.*, 1982). Finally, reference will be made to the pleomorphic large spherical (1.5–2.5 μm in diameter), Gram-positive motile cells of the genus *Methanosarcina* (G + C ratio of the DNA = 39–51 moles %). Of the two species, i.e. *M. barkeri* and *M. methanica*, found in the sea, the latter has never been isolated in pure culture, and for that matter, is not included in the *Approved Lists of Bacterial Names* (Skerman *et al.*, 1980) or their supplements.

4.1.12   *Aerobic Gram-positive cocci*

There has been sporadic reference made to the presence of aerobic Gram-positive cocci in the marine environment. At various times, these organisms have been identified as *Marinococcus* (formerly known as *Planococcus*), *Micrococcus* and *Staphylococcus*. *Marinococcus* was proposed to accommodate motile cocci, which contained meso-diaminopimelic acid in the cell wall (Van Hao *et al.*, 1984). These strictly aerobic motile (one or two flagella) organisms produce yellow or white colonies on laboratory media such as marine 2216E agar, and are catalase-positive. Isolates (from seawater) have been classified in two species, as *M. albus* and *M. halophilus*, with the G + C ratio of the DNA reported to be 43.9–46.6 moles %.

Anderson (1962) reported the presence of Gram-positive cocci (micrococci/staphylococci) in the North Sea. However, of 205 isolates, species rank was not assigned. In a subsequent investigation Gunn and Colwell (1983) executed a numerical taxonomy study on 220 aerobic, catalase-positive Gram-positive cocci which were derived from marine and estuarine surface waters. From the computer analyses, 86% of the isolates were equated with *S. epidermidis* or *S. hominis*, 5% with *S. aureus* and 9% with *Micrococcus* spp. Nevertheless, it should be emphasised that such organisms account for only a small proportion of the total marine aerobic heterotrophic bacterial community.

4.1.13   *Gram-positive endospore-forming rods*

It is a moot point whether *Bacillus* and *Clostridium* species constitute active components of the marine microflora or inactive (as spores) washed-in propagules from the terrestrial environment. Yet within marine sediments, representatives of both genera may be found. Indeed, species have been proposed to accommodate marine isolates. For example, Rüger (1983) described a marine *Bacillus* i.e. *B. marinus* (this organism was previously referred to as *B. globisporus* subsp. *marinus*), which demonstrated a pronounced requirement for sodium chloride. The organism grew at 5–30 °C in aerobic conditions, was motile by peritrichous flagella, produced catalase but not arginine dihydrolase, ß-galactosidase, indole or lysine or ornithine decarboxylase, degraded aesculin but not lipids or urea, and was negative for the methyl red test and the Voges Proskauer reaction. The G + C ratio of the DNA was considered to be approximately 39 moles %. Other *Bacillus* isolates were studied by Boeyé and Aerts (1976) but little attempt was made at identification/classification to species level. Regarding their anaerobic counterparts, i.e. *Clostridium*, marine isolates have been classified in a novel taxon, namely *C. oceanicum* (Smith, 1970; Gottschalk *et al.*, 1981). Isolates grow optimally at 30–37 °C, tolerate up to 4% (w/v) sodium chloride, degrade lecithin, and are proteolytic and saccharolytic (slightly). The G + C ratio of the DNA of this organism has been calculated as 27–28 moles %. Undoubtedly, other clostridia occur in marine sediments, but there have been difficulties with speciation due to the general paucity of information.

4.1.14 *Actinomycetes and related bacteria*

Although fairly uncommon in the marine environment, actinomycetes and related organisms do occur. Facultative or strictly anaerobic Gram-positive actinomycetes which do not form aerial mycelia or spores have been equated with *Actinomyces*. However there is no obvious species grouping for these isolates, which occur principally in sediments (Schaal and Pulverer, 1981). Similarly there is confusion over the precise identity of actinomycetes with single spores but without aerial hyphae, which degrade complex molecules. Such isolates appear to belong in the genus *Micromonospora*. These organisms may be recovered from beaches and coastal waters and sediment (Cross, 1981). Gram-positive mycelial actinomycetes, regarded as *Streptomyces* spp., also occur in sediments but again the taxonomy is unsettled (Kutzner, 1981).

Aerobic Gram-positive acid- to partially acid-fast mycolic acid-containing actinomycetes with chemo-type IV walls, which produce primary but not aerial mycelia and fragment into bacillary and coccoid elements occur quite commonly in coastal marine locations. These organisms, with a G + C ratio of the DNA of 64–69 moles %, possess the key characteristics of *Nocardia* and *Rhodococcus* (Goodfellow and Minnikin, 1981). Although most isolates remain unspeciated some have been identified as *N. marina*, a species of dubious taxonomic validity. Rather more is known about marine representatives of *Rhodococcus*, which have been grouped in three species, namely *R. marinonascens* (Helmke and Weyland, 1984), *R. maris* (Nesterenko *et al.*, 1982) and *R. rhodochrous* (Helmke and Weyland, 1984). These are non-motile, salt-requiring, red, orange or cream-pigmented organisms which produce catalase, acid from *D*-fructose, *D*-glucose and mannose, rarely reduce nitrate and mostly degrade aesculin and tyrosine, but not casein, chitin, DNA, gelatin, starch, xanthine or urea. The G + C ratio of the DNA ranges from 64.9 to 73.2 moles %. Generally, the organisms seem to be restricted to the uppermost layers of sediment (Weyland, 1969, 1981a, b; Weyland *et al.*, 1970).

Weakly Gram-positive, acid-fast rods containing mycolic acids have been identified as *Mycobacterium* spp. (Kubica and Good, 1981). These organisms may be notoriously difficult to isolate, possibly reflecting complex, poorly understood nutritional requirements. Representatives may be pathogenic, e.g. *M. marinum* on fish, or saprophytic in sediments and on plants and animals.

Difficult to identify pigmented (often orange), Gram-positive rods, with or without a so-called coryneform morphology, have been labelled as coryneforms. The genus groupings probably include *Arthrobacter, Brevibac-*

terium (= *B. linens*), *Corynebacterium* and *Curtobacterium*. These organisms occur in seawater and on fish.

4.1.15   *The spirochaetes*

Helical Gram-negative bacteria 0.4 μm × 10–15 μm in size, which are motile by two or more flagella, abound in anoxic marine muds. It appears that the diamino acid in the peptidoglycan is *L*-ornithine. *Crispispira pectinis*, with its bundles of flagella, occurs in marine molluscs. However the organism has not been grown in pure culture (Canale-Parola, 1984). Therefore its taxonomic status is highly questionable. At least the ability to isolate several of the marine *Spirochaeta* species (*S. isovalerica* and *S. litoralis*) have aided their taxonomy. Thus the obligate anaerobic *S. litoralis* may be cultured on chemically defined low-nutrient media using rifampicin as a selective agent. In contrast, *S. plicatilis*, considered to be found in association with *Beggiatoa*, has not yet been obtained in pure culture. *S. litoralis* is a non-pigmented catalase-negative, salt-requiring organism, which does not reduce nitrates, with a temperature range of growth of 15–35 °C. The G + C ratio of the DNA of *S. litoralis* is approximately 51 moles %, whereas *S. isovalerica* is somewhat higher at 63.6–65.6 moles %. This difference implies marked generic heterogeneity. This latter taxon is differentiated from other marine spirochaetes by the formation of 4- and 5-carbon branched fatty acids, i.e. isovalerate, 2-methylbutyrate and isobutyrate, as fermentation products (Harwood and Canale-Parola, 1983). The four strains of *S. isovalerica* examined were recovered from intertidal muds, whereas *S. litoralis* is common in sulphite-containing muds.

4.2   Eukaryotes

4.2.1   *Micro-algae*

4.2.1.1   *Diatoms* Diatoms are photosynthetic algae, which are characterised by an ornate siliceous shell (frustule), comprising the valva (valve) and pleura (girdle) (Fig. 4.2). There are two long narrow slits, referred to as raphes, to which there is a correlation with motility (McIntire and Moore, 1977). Many thousands of species of marine diatoms have been described, of which the taxonomy is based on the morphology of the frustule (see Table 4.2 for an outline classification). Diatoms belong to the class Bacillariophyceae. However, the division into orders is unsettled. Yellow or brown chloroplasts are exhibited intracellularly. Planktonic forms may be appendaged, e.g. the bloom-forming *Chaetoceros*.

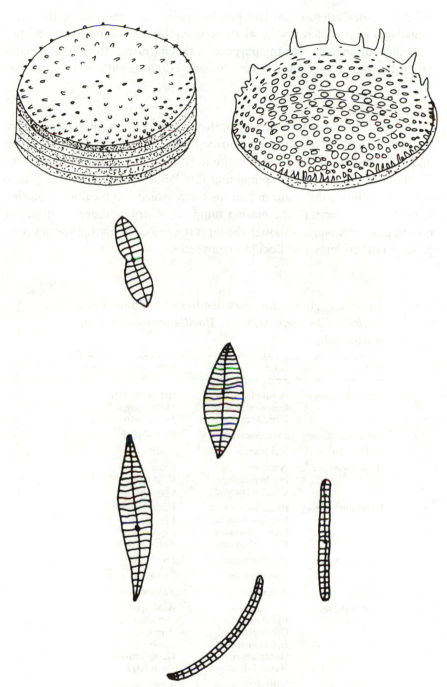

Fig. 4.2 Examples of the morphology of marine diatoms.

4.2.1.2   Other micro-algae The precise taxonomic position of the dino-flagellates is unsettled insofar as there is controversy about whether they are plants or animals. For the purpose of this narrative, the dinoflagellates and other phyto-flagellates will be considered along with other protozoa.

4.2.2   *Filamentous fungi*

According to Kohlmeyer and Kohlmeyer (1979), there are approx-imately 300 species of obligately marine fungi, of which the overwhelming majority are Ascomycotina. Approximately one-third of the total comprise Deuteromycotina, with the remaining few being Basidiomycotina (Table 4.3). In addition, there are numerous taxa, which are facultative marine inhabitants. In general, the marine fungi are distributed across all major mycological groupings, although the largest number of marine species com-prise Pyrenomycetes and Loculoascomycetes.

Table 4.2. *An outline classification of diatoms, i.e. Division Chrysophyta, Class Bacillariophyceae, Order Bacillariales*

Suborder	Family	Representative genus
Coscinodiscineae	Actinodiscaceae	*Actinoptychus*
	Coscinodiscaceae	*Skeletonema*
	Hemidiscaceae	*Hemidiscus*
Aulacodiscineae	Eupodiscaceae	*Eupodiscus*
Auliscineae	Auliscaceae	*Auliscus*
Biddulphiaceae	Anaulaceae	*Anaulus*
	Biddulphiaceae	*Biddulphia*
	Chaetoceraceae	*Chaetoceros*
Rhizosoleniineae	Bacteriastraceae	*Bacteriastrum*
	Corethronaceae	*Corethron*
	Leptocylindraceae	*Leptocylindrus*
	Rhizosoleniaceae	*Rhizosolenia*
Fragilariineae	Fragilariaceae	*Fragilaria*
Eunotiineae	Eunotiaceae	*Eunotia*
Achnanthineae	Achnanthaceae	*Achnanthes*
Naviculineae	Auriculaceae	*Auricula*
	Bacillariaceae	*Nitzschia*
	Cymbellaceae	*Amphora*
	Epithemiaceae	*Epithemia*
	Gomphonemaceae	*Gomphonema*
	Naviculaceae	*Pinnularia*
Surirellineae	Surirellaceae	*Surirella*

After Hendey (1964).

Table 4.3. *Filamentous fungi recovered from the marine environment*

Sub-division	Class	Family	Genus
Ascomycotina	Discomycetes	Orbiliaceae	*Orbilia*
	Plectomycetes	Eurotiaceae	*Amylocarpus, Eiona*
	Pyrenomycetes	Laboulbeniaceae	*Laboulbenia*
		Physosporellaceae	*Buergenerula*
		Spathulosporaceae	*Spathulospora*
		Diaporthaceae	*Gnomonia*
		Halosphaeriaceae	*Aniptodera, Bathyascus, Carbosphaerella, Ceriosporopsis, Chadefaudia, Corollospora, Haligena, Halosarpheia, Halosphaeria, Lignincola, Lindra, Lulworthia, Nais, Nautosphaeria, Trailia.*
		Hypocreaceae	*Halonectria, Heleococcum, Hydronectria, Nectriella*
		Polystigmataceae	*Haloguignardia, Phycomelaina*
		Sordariaceae	*Biconiosporella, Zopfiella*
		Sphaeriaceae	*Abyssomyces, Chaetosphaeria, Pontogeneia*
		Verrucariaceae	*Pharcidia, Turgidosculum*
		Incertae sedis	*Oceanitis, Ophiobolus, Savoryella, Torpedospora*
	Loculoascomycetes	Herpotrichiellaceae	*Herpotrichiella*
		Mycoporaceae	*Leiophloea*
		Mycosphaerellaceae	*Didymella, Mycosphaerella*
		Patellariaceae	*Bankegyia, Kymadiscus*
		Pleosporaceae	*Didymosphaeria, Halotthia, Helicascus, Keissleriella Leptosphaeria, Manglicola, Massarina, Microthelia, Paraliomyces, Phaeosphaeria, Pleospora, Pontoporeia, Thalassoascus, Trematosphaeria*
		Incertae sedis	*Crinigera, Orcadia, Sphaerulina*
Basidiomycotina	Gasteromycetes	Melanogastraceae	*Nia*
	Hymenomycetes	Corticiaceae	*Digitatispora*
		Incertae sedis	*Halocyphina*
	Teliomycetes	Tilletiaceae	*Melanotaenium*
Deuteromycotina	Hyphomycetes	Agronomycetaceae	*Papulaspora,*
		Moniliaceae	*Blodgettia, Botryophilalophora, Clavatospora, Varicosporina*
		Dematiaceae	*Asteromyces, Cirrenalia, Cladosporium, Clavariopsis, Cremasteria, Dendryphiella, Dictoyosporium, Drechslera, Humicola, Monodictys, Orbimyces, Periconia, Sporidesmium, Stemphylium, Trichocladium, Zalerion*
		Tuberculariaceae	*Allescheriella, Tubercularia*
	Coelomycetes	Excipulaceae	*Dinemasporium*
		Melanconiaceae	*Sphaceloma*
		Sphaerioidaceae	*Ascochyta, Ascochytula, Camarosporium, Coniothyrium, Cytospora, Diplodia, Macrophoma, Phialophorophoma, Phoma, Rhabdospora, Robillarda, Septoria, Stagonospora*

Based on Kohlmeyer and Kohlmeyer (1979); Kohlmeyer (1986).

4.2.3   *Yeasts*

A total of 177 species of yeasts (unicellular fungi) have been recovered from the marine environment, of which the minority are truly obligate
(Kohlmeyer and Kohlmeyer, 1979) (Table 4.4).

4.2.4   *Protozoa*

The phylum Protozoa encompasses the simplest forms (unicellular)
of animals. A simplified classification, based on Meglitsch (1972), has been
included in Table 4.5. Within the marine environment, there are many
thousands of species. Representatives of the order Dinoflagellida (Table
4.6; Figs. 4.3, 4.4) comprise the predominant component of the zooplankton. For example, *Noctiluca scintillans* generate phosphorescence in
the sea due to a luciferin–luciferase system, whereas *Gonyaulax tamarensis*
and *Gymnodinium breve* cause the red tides. In contrast, the order
Radiolarida and Foraminiferida contain representatives which are common
marine pelagic and benthic forms, respectively. Ciliates are especially common in shallow benthic habitats and marine sands (Sleigh, 1973). Amoebae
are quite rare.

Dinoflagellates are unicellular organisms, each of which possesses a large
nucleus with chromosomes, two dissimilar flagella and one or more yellow-
brown coloured chloroplasts containing chlorophyll *a* and *c*, β-carotene,
and either peridinin or fucoxanthin. There may be a cell covering of thecal
plates. Trichocysts may be present. These are ejected from the cell during
adverse conditions. Most cells possess a pair of elaborately structured
'pusules' which are associated with the flagella pores, presumably function-

Table 4.4. *Marine yeasts*

Sub-division	Family	Genus
Ascomycotina	Metschnikowiaceae	*Metschnikowia*[a]
	Saccharomycetaceae	*Debaryomyces, Hanseniaspora, Hansenula, Kluyveromyces*[a], *Pichia*[a], *Saccharomyces*
Basidiomycotina	Sporobolomycetaceae	*Leucosporidium*[a], *Rhodosporidium*[a], *Sporobolomyces*
Deuteromycotina	Torulopsidaceae	*Candida*[a], *Cryptococcus*[a], *Kloeckera, Rhodotorula*[a], *Sterigmatomyces*[a], *Sympodiomyces*[a], *Torulopsis*[a], *Trichosporon*
	(Yeast-like cells)	*Aureobasidium/Pullularia*

[a]Contains obligate marine species.
After Kohlmeyer and Kohlmeyer (1979).

Table 4.5 *An outline classification of the phylum Protozoa*

Sub-phylum	Super class	Class/sub-class	Order
Sarcomastigophora (unciliated; possess flagella or pseudopodia)	Mastigophora (flagellate)	—	Chloromonadida (possesses green chromoplasts, 2 flagella-one trailing)
			Choanoflagellida (possesses 1 flagellum)
			Chrysomonadida (possess chromoplasts)
			Cryptomonadida (possess chromoplasts)
			Dinoflagellida (possess one transverse and one trailing flagella)
			Diplomonadida (parasitic)
			Euglenida (possess green chromoplasts)
			Hypermastigida
			Kinetoplastida (1–4 flagella; parasitic)
			Retortamonadida (1–4 flagella; parasitic)
			Rhizomatigida (pseudopodia; 1–4 flagella)
			Trichomonadida (parasitic)
			Volvocida (possess 2–4 apical flagella; 2 chromoplasts)
	Opalinata (ciliate; parasitic)	—	Opalinida
	Sarcodina (pseudopodia; no flagella on adult)	Rhizopodea	Amoebida (naked cells)
			Arcellinida (rigid cell)
		Filosia (tapering filopodia)	
		Granuloreticulosia (delicate reticulopodia)	Foraminiferida (pseudopodia; cell enclosed in chambered chitin envelope)
		Mycetazoia (slime molds)	Acrasida
		Labyrinthulida (forms slime net)	
		Actinopodea (floating cells with radiating pseudopodia)	Acantharida (marine)
			Heliozoida (freshwater)
			Radiolarida (marine)
		Piroplasmea (parasites)	

Table 4.5 (cont.)

Sub-phylum	Super class	Class/sub-class	Order
Ciliophora (ciliated protozoa)	Ciliatea	Holotrichia	Apostomatida Astomatida (parasitic in oligochaetes) Chonotrichida (sessile) Gymnostomatida Hymenostomatida Thigmotrichida (present in bivalves) Trichostomatida
		Peritrichia (sessile; possess stalks)	Peritrichida
		Suctoria (sessile; unciliated adults)	Suctorida
		Spirotrichia	Entodiniomorphida Heterotrichida Hypotrichida Odontostomatida Oligotrichida (marine) Tintinnida (marine)
Sporozoa (spore-forming parasites; unciliated, unflagellated)	Telosporea (gliding movement)	Gregarinia (large trophozoites)	Archigregarinida Eugregarinida Neogregarinida
		Coccidia (small trophozoites)	Eucoccida Protococcida
	Toxoplasmea (sporozoa not forming spores) Haplosporea (sporozoa form spores)		
Cnidospora (parasitic; form spores)	Myxosporidea (spores formed from multi-cellular sporoblast)	—	Actinomyxida Microsporidea Myxosporida

After Meglitsch (1972).

ing in osmoregulation. The organisms reproduce both asexually, by binary fission, and sexually. Life cycles are exhibited, and cysts are produced (Dodge, 1982). In the marine environment, most dinoflagellates are free-living. Classification is based overwhelmingly on morphology, namely overall shape and position of the girdle. Zoologists regard the organisms as animals, and follow the edicts of the International Code of Zoological Nomenclature. By this system, dinoflagellates are classified in the order Dinoflagellida of the super-class Mastigophora. However, botanists have regarded dinoflagellates as plants, due to the presence of chloroplasts. Consequently, botanists have followed the terms of the International Code of Botanical Nomenclature, with the dinoflagellates classified as Dynophyta.

Table 4.6. *An outline classification of the dinoflagellates*

Order	Family	Representative genus
Procentrales (anteriorly inserted flagella; 2 main thecal plates)	Procentraceae	*Procentrum*
Dinophysiales (flagella inserted near anterior end; transverse groove present; thecal plates present)	Amphisoleniaceae Dinophysiaceae	*Amphisolenia Dinophysis*
Gymnodiniales (naked cells, i.e. no substantial thecal plates)	Gymnodiniaceae Lophodiniaceae Polykrikaceae Pronoctilucaceae Warnowiaceae	*Gymnodinium Aureodinium Polykrikos Pronoctiluca Warnowia*
Noctilucales (large naked cells often with single tentacles; complex reproduction)	Noctilucaceae	*Noctiluca*
Pyrocystales (non-motile coccoid cells; reproduction is by a motile gonyaulacoid/ gymnodinioid cell)	Pyrocystaceae	*Pyrocystis*
Peridiniales (distinct thecal plates resembling armour; transverse groove present)	Cladopyxidaceae Gonyaulacaceae Heterodiniaceae Oxytoxaceae Peridiniaceae Podolampaceae Pyrophacaceae Triadiniaceae	*Micracanthodinium Gonyaulax Heterodinium Oxytoxum Protoperidinium Podolampas Pyrophacus Triadinium*
Blastodiniales (parasitic; reproduction is by a motile gymnodinioid cell)	Dissodiniaceae	*Dissodinium*

After Dodge (1982).

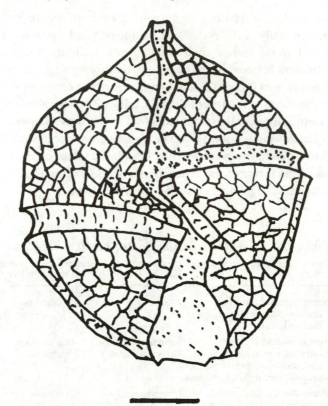

Fig. 4.3 The marine dinoflagellate, *Gonyaulax*. Bar equals 10 μm.

Ciliates are characterised by the dominant covering of cilia. The organisms possess two types of nuclei (referred to as macro- and micro-nuclei), and reproduce both sexually and by binary fission (Sleigh, 1973). Many taxa have been recovered from the marine environment. For example, in shallow benthic habitats representatives of Heterotrichida (*Folliculina aculeata; Gruberia lanceolata*; Fig. 4.5), Hymenostomatida (*Cohnilembus verminus; Uronema marina*; Fig. 4.6), Hypotrichida (*Aspidisca steini; Diophrys appendiculata*) and Gymnostomatida (*Mesodinium pulex*; Fig. 4.7) abound. In marine sands Hymenostomatida (*Pleuronema coronatum*; Fig. 4.8) and Gymnostomatida (*Centrophorella lanceolata; Chaenia gigas; Geleia fossata*; Fig. 4.9; *Lionotus* spp.; *Remanella rugosa; Trachelocerca phoenicopterus*) may be found.

Finally, reference will be made to the foraminiferans (Figs. 4.10–4.13), which occur in large numbers on the sea floor. These protozoans build shells containing calcareous material, which upon death of the organisms, contribute to fossil deposits.

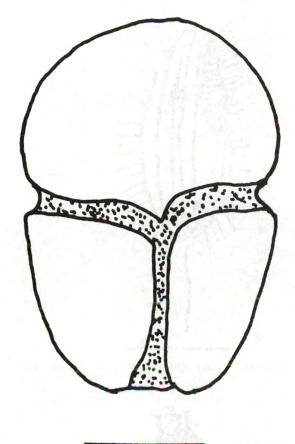

Fig. 4.4 *Gymnodium*. Bar equals 10 μm.

Fig. 4.5 *Gruberia lanceolata*. Bar equals
500 μm (after Sleigh, 1973).

Fig. 4.6 *Uronema marina*. Bar equals 40 μm (after Sleigh, 1973).

Fig. 4.7 *Mesodinium pulex*. Bar equals 25 μm (after Sleigh, 1973).

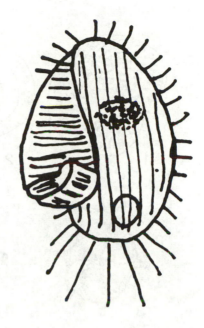

Fig. 4.8 *Pleuronema coronatum*. Bar equals 80 μm (after Sleigh, 1973).

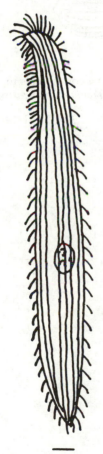

Fig. 4.9 *Geleia fossata*. Bar equals 350 μm (after Sleigh, 1973).

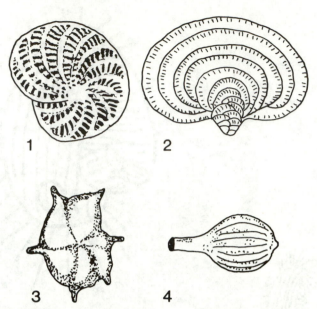

Fig. 4.10 Foraminiferan morphology. 1, 2, 3 and 4 corresponds to *Elphidium crispa, Pavonina flabelliformis, Hantkenia alabamensis* and *Lagena striata*, respectively.

5   Ecology

It is apparent that the numbers and taxonomic composition of the marine microbial population are greatly influenced by localised changes in nutrient supply, availability of oxygen or hydrogen sulphide, and temperature. Away from the coast, with its enhanced nutrient supply through terrestrial runoff and pockets of pollution, the majority of the seas are lacking in nutrients. In particular, organic nutrients are in short supply. Yet apart from the photosynthetic organisms in the photic zone, a diverse range of microbial heterotrophs abounds, albeit often in low numbers. The topic of survival of organisms in the sea has prompted some excellent work.

5.1   Survival of indigenous organisms in the marine environment

The approach to microbial ecology of counting bacterial colonies on rich laboratory media is hardly conducive to sound scientific practice. Yet much of the earlier (and unfortunately some more recent) work centred on this theme. So, from colonies on an agar medium incorporating high levels of organic nutrients, marine bacteria were inevitably observed as large (often pleomorphic) rods and cocci. Direct microscopic examination of fresh marine samples would tend not to support this contention. Indeed, many prokaryotes, and for that matter eukaryotes, appear to be very much smaller, when examined *in situ*. Viable bacterial cells are able to pass through the pores of 0.45 µm diameter bacteriological filters (Anderson and Heffernan, 1965). By phase contrast microscopy, such small cells often appear to be highly motile. Questions to be posed concern the role of these small cells – do they constitute the normal size of marine bacteria; are they part of a life cycle; are they starved; could they form a survival stage; and are they active or dormant?

Kogure *et al.* (1979, 1980) and Simidu *et al.* (1983) highlighted the differences between results obtained with direct counts as opposed to total viable counts. Assuming that all the structures observed by microscopy were microorganisms, namely bacteria (this is far from proven), then <0.1% formed colonies on agar media. However, by use of a modified microscopic technique involving nalidixic acid, which suppresses cell division, a direct viable count was obtained (Table 3.1). Again, the values were much less than the direct count but higher than the total count. The difference in values suggests that a sizeable proportion of the microbial population is non-culturable (at least with regard to the techniques used) but not necessarily non-viable. In a separate study, Torrella and Morita (1981) used phase contrast microscopy coupled with a micro-culture method to examine the response of marine bacteria to high levels of nutrient, i.e. higher concentrations than normally associated with seawater. Essentially, the response could be described as either a rapid increase in size commensurate with a high multiplication rate (these cells were likened to the concept of a zymogenous flora) or a much slower growth rate with the development of micro-colonies. The latter reaction occurred with very small coccoid bacteria of only 0.3 μm in diameter (these are the so-called ultra-micro-bacteria). These were considered as autochthonous, i.e. the true marine bacteria. Obviously with such organisms, there would not be a rapid multiplication following the availability of abundant organic nutrients. Indeed, high levels of such nutrients may even be debilitory to the organisms. It is predicted that these organisms will be admirably suited to scavenge for the normally low levels of nutrients associated with seawater. Indeed, Carlucci and co-workers (e.g. Carlucci and Shimp, 1974) recovered Gram-negative motile rods which grew in low levels of nutrient, such as in unsupplemented seawater. Yet other workers have described changes in bacteria which appear to enable them to survive starvation.

Novitsky and Morita (1976, 1977, 1978) isolated a psychrophilic *Vibrio*, coined as Ant-300, from Antarctic Convergence waters in 1972. Upon starvation, this strain increased initially in numbers by up to 200-fold, and became much smaller, i.e. forming 'ultra-micro-cells'. An initial explanation was that the increase in cell numbers reflected a survival mechanism. It should be emphasised that Antarctic Convergence waters mix and sink, whereupon they do not appear on the surface for several centuries. With the addition of nutrients, and thus recovery from starvation, there was a marked increase in cell size before division, indicating that a minimal size must be reached prior to binary fission proceeding (Novitsky and Morita, 1976). With starvation of Ant-300 there was a rapid decline in the levels of DNA, RNA and specific proteins to a constant low level (Amy *et al.*, 1983a; Amy and Morita, 1983). Interestingly, the RNA decreased initially but then increased. Yet after six weeks of starvation, 45–60% of the cells were capable of respiring, as determined by the reduction of tetrazolium to coloured formazan (Amy *et al.*, 1983a). Upon addition of nutrients, there was an apparent lag period before a response could be detected (Amy and Morita, 1983). During this period, there was an increase in DNA, RNA and protein before division (Amy *et al.*, 1983b). Could the starvation–survival (this term was coined by Morita, 1982) of Ant-300, with the development of very small cells, explain the abundance of 'mini-bacteria' in the water column? An assumption could be made that all such small cells are staving off the effects of starvation. Additional evidence has been provided by Dawson *et al.* (1981) and Kjelleberg *et al.* (1982). In particular, the latter team reported an increase in overall number commensurate with a decline in cell size of a marine *Vibrio* at the onset of starvation. Moreover, it was concluded that better survival occurred at interfaces. Here, cells were notably smaller, i.e. 0.32 μm^3, within 22 hours of initiating starvation, than bacteria in an aqueous phase (cell volume = 0.42 μm^3). Subsequently, it was reported that new proteins may be sythesised during starvation (Jaan *et al.*, 1986); a rather intriguing feat. In addition, substantive chemical changes may occur in the cell envelope (Malmcrona-Friberg *et al.*, 1986). In all cases of starvation, maintenance energy must inevitably be derived from intracellular constituents; hence the decline in amounts of macromolecules. Using a chemostat, Jones and Rhodes-Roberts (1981) calculated the maintenance energy of a marine *Pseudomonas* as 0.042 g glucose/g dry weight/h.

If many marine bacteria are in a starvation–survival situation, it is relevant to enquire about whether such cells should be considered as active or dormant. The ability of Morita to maintain starved cells of Ant-300 for approximately 2.5 years must surely cast doubt about whether the cells should be regarded as active. Therefore to extend the analogy to natural marine microbial communities, it may be tempting to speculate that many cells may not be active, but nevertheless still viable.

Stevenson (1978) broached the topic of dormancy to explain the ability of bacteria to survive in the aquatic environment. According to Sussman and Halvorson (1966), dormancy was defined as: 'any rest period or reversible interruption of the phenotypic development of an organism'. Dormancy may be 'constitutive', as for example with endospore production by *Bacillus*, or 'exogenous'. Stevenson defined exogenous dormancy as: 'a condition in which development is delayed because of unfavourable chemical or physical conditions of the environment'. It is unlikely that scientists would contend the notion of constitutive dormancy. The phenomenon may well explain the survival of *Bacillus* spp., as endospores in marine sediments. However, the possibility of exogenous dormancy, in which to voice Colwell's analogy, the vegetative cell is akin to a spore without its heat-resistant wall, is a situation which many scientists find difficult to accept. If a school of thought defines viability (of bacteria) in terms of plate counts, then a notion of 'non-culturable but viable' would be difficult to grasp. In his landmark publication, Stevenson (1978) cited several examples to support the concept of exogenous dormancy. Jannasch (1967), using continuous culture techniques, showed that marine bacteria would not divide in seawater which contained limiting substrate concentrations. The suggestion was made that the quantity of nutrients was below the threshold level necessary for cell division. Therefore, there were cells surviving, albeit inactive. In another example, Williams and Gray (1970) mentioned two distinct increases in the rates of uptake of amino acids by marine populations. The first increase occurred immediately after the addition of substrate, whereas the second response was noted after 20–36 h. This second increase was attributed to a possible re-activation of inactive cells or to microbial growth. Vaccaro and Jannasch (1966) determined that natural bacterial communities in seawater reacted rapidly to enrichment with organic nutrients. Thus there was a good kinetic response within 12 h, which Wright (1973) considered to mean that some cells in the community were functionally switched off or dormant. This explanation was forthcoming because the 12 h response seemed unlikely to allow time for substantial multiplication to occur. Subsequently, Colwell and co-workers (Xu *et al.*, 1982) demonstrated survival of apparently non-culturable cells of *Vibrio cholerae* in seawater. Here, methods centred upon the nalidixic acid reaction (Kogure *et al.*, 1979) and epifluorescence microscopy. By use of low concentrations of nalidixic acid together with organic nutrients, which stimulated metabolism, Xu *et al.* (1982) observed that coccoid cells of *V. cholerae* were capable of enlarging and metabolising substrates. Certainly the question as to whether or not cells observed by microscopy are truly viable has been resolved by evidence obtained by epifluorescence microscopy, which measures the integrity of the bacterial outer membrane (Costerton *et al.*, 1974; Coleman and Leive, 1979), and by uptake of ^{14}C-labelled substrates, as shown by Singleton *et al.* (1982) for *V. cholerae*. In short, fluorescing cells, as observed by epifluorescence microscopy, which do not produce colonies on agar-containing media, are judged to be intact and viable.

Undoubtedly, the concept of dormancy in relation to marine microbial populations will be heatedly debated before final acceptance as an established scientific fact.

5.2   Fate of non-indigenous organisms in the marine environment

There may be justifiable concern about the ability (or inability) to distinguish indigenous from non-indigenous organisms in the marine environment. To extend a previously mentioned scenario, it may be expected that non-indigenous organisms may quickly die, enter a dormant (non-culturable) phase, or become established as part of the marine microflora for protracted or prolonged periods. Work has highlighted the fate of pathogens in marine habitats. Certainly it is accepted that some organisms are not killed upon entering the marine environment. Instead, the cells become debilitated and/or dormant. For example, some coliforms are not killed in seawater, but gentle techniques are required for their subsequent recovery and growth in laboratory media (see Dawe and Penrose, 1978; Olson, 1978). Other workers have demonstrated the presence of dormant/non-culturable cells, notably of *Vibrio cholerae* (Xu *et al.*, 1982; Colwell *et al.*, 1985). Still other non-indigenous organisms may be attacked by members of the resident marine microflora.

Living cells, such as *Escherichia coli*, may be preyed upon by other micro-organisms, notably *Bdellovibrio, Flexibacter, Polyangium* and amoebae, e.g. *Vexillifera* (Mitchell, 1971; Enzinger and Cooper, 1976; Roper and Marshall, 1978). With the presence of prey, the predators increase in number before returning to the former population level. Usually, this occurs as a result of large protozoa which consume the predators. With fungi there may be a rapid reduction in the amount of biomass to a low level, which is able to survive for long periods in the sea. An example is the situation with *Candida albicans* (Jamieson *et al.*, 1976). Other fungi may be actively lysed as a result of bacterial extracellular enzymes. Thus Mitchell and Wirsen (1968) observed lysis of the fungal cell wall in non-marine oomycetes, namely *Achlya, Apodachlya, Isoachlya, Saprolegnia* and *Thraustotheca*. Here lysis appeared to be most marked at alkaline pH, i.e. pH 8.0, and in the presence of organic nutrients such as yeast extract.

When microbial cells die there is every indication that there will be rapid degradation, such that dead organisms are quickly removed. Using beach sand Novitsky (1986) recorded that RNA was degraded more rapidly than DNA, although both nucleic acids were eventually decomposed to a similar extent, i.e. 60–70%. In addition, 32% of the carbon incorporated into macro-molecules was degraded/respired within 10 days.

5.3   Predator–prey relationships (food webs)

The involvement of marine micro-organisms with the destruction of non-indigenous microbes has already been discussed. However, predation is not purely directed against alien cells in the sea. Indeed it is a commonly occurring process between many components of the marine microflora (e.g. Goldman and Caron, 1985; Goldman *et al.*, 1985). Bacteria, e.g. *Bdellovibrio*, lyse other bacteria; small protozoa prey upon bacteria and algae, e.g. diatoms (e.g. Caron *et al.*, 1986); large protozoa attack smaller members of the same zoological kingdom; animals, e.g. whales, consume massive quantities of plankton.

It has been estimated that between 30 and 50% of primary productivity in the marine environment is channeled through heterotrophic bacteria (Williams, 1981; Azam *et al.*, 1983; Ducklow, 1983; Hagström, 1984). Thus, it would seem that there is a large quantity of bacterial biomass available to the food web. For example, holothurians in sediment have been credited with grazing bacteria to the extent that one particular organism, *Holothuria atra*, obtained 10–40% of its carbon from bacteria during each summer day (Moriarty *et al.*, 1985). Experimental evidence has tended to be derived from laboratory rather than *in situ* work. It is interesting that both protozoa (Fenchel, 1982, 1986; Sherr *et al.*, 1983) and polychaetes (Montagna, 1984) have been cited as principal predators of bacteria. Indeed Fenchel (1986) reported that microflagellates clear 30–50% of marine bacteria per day. In coastal waters at Limfjorden, Denmark, heterotrophic nanoplankton were considered to clear approximately 45% of the bacteria each day (Andersen and Sørensen, 1986). Meanwhile, ciliates cleared approximately 93% of the heterotrophic nanoplankton during a similar time period (Andersen and Sørensen, 1986). It was calculated by Campbell and Carpenter (1986) that at one site in the Gulf of Maine, 37–57% of the biomass was consumed by grazers.

Using a model system with genetically marked minicells of *Escherichia coli* and the flagellate, *Ochromonas*, Wikner *et al.* (1986) noted that the rate of disappearance of the bacteria was initially low, i.e. 4×10^5 minicells/ml/h at 3 h, but increased commensurate with the growth of the predator. In fact, when grown on minicells, the growth rate of *Ochromonas* was calculated to be 43%. The fraction of bacterial biomass transferred to the flagellate was in the region of 27–111%. Interestingly, similar results were obtained with native marine bacteria. Wright and Coffin (1984) published data which showed that in coastal water with total bacterial populations of 2×10^6–3×10^6/ml, daily grazing accounted for approximately 1×10^6 bacteria/ml. This suggested that predation reduced the populations by approximately 33–50%.

However only 10–15% of the biomass from each trophic level is transferred to the next step. The majority, 85–90%, of the biomass fuels respiration. In the case of phytoplankton, it would appear that 75% of the energy is directed towards so-called 'grazing' food chains, whereas the remaining amount goes to 'decomposer' food chains (Stockner and Antia, 1986).

Fuhrman and McManus (1984) reported the presence of small, as yet unseen and unidentified eukaryotes, which were possibly responsible for the grazing of >50% of bacteria in coastal water. Such eukaryotes, which apparently were capable of passing through 0.6 μm pore filters, were studied by means of eukaryotic inhibitors, i.e. cycloheximide. Whether or not these eukaryotes were small protozoa is the subject of conjecture. Yet small (2–10 μm in diameter) protozoa have been attributed with an important ability to graze bacteria (Fuhrman and McManus, 1984). Indeed, Caron *et al.* (1982) demonstrated the importance of macro-aggregates (marine snow) in food chains. In this landmark study, both bacteria and bacteriovorous protozoa were more common, i.e. by 10–10 000 times, in marine snow than in the surrounding water.

An assumption may now have been reached that the principal grazers are protozoa. However this is far from true. For example, it has been noted that nematodes grow well on both bacteria (Schiemer *et al.*, 1980; Schiemer, 1982) and diatoms (Alongi and Tietjen, 1980). Similarly harpacticoid copepods graze upon bacteria (Rieper, 1978) and diatoms (Ustach, 1982). Moreover in an excellent article, Montagna (1984) reported on the grazing of bacteria and diatoms by copepods, nematodes, ostracods and polychaetes (Table 5.1). By studying marine sediments from the US Atlantic seaboard, Montagna (1984) deduced that the predominant grazers were, in fact, polychaetes (Table 5.1). On a quantitative basis, polychaetes exerted a greater effect upon bacterial and diatom populations than the combined activity of all the other grazers. Although microbial activity was greater in summer than winter, there were negligible differences in grazing rates between the two seasons. Overall the results of Montagna indicated that 3% of bacteria and 1% of diatoms were removed hourly by grazing. Obviously this has a dramatic effect on marine microbial populations.

Radioactive tracers, e.g. ^{14}C-bicarbonate, feature prominently in most studies of food chains (e.g. Conover and Francis, 1973; Daro, 1978). Grazing rates may be ascertained from mathematical models. Montagna (1984) used

Table 5.1 *Grazing rates of marine bacteria and diatoms by copepods, ostracods, nematodes and polychaetes during summer*

Predator	Grazing rates (x 10^3/h)		Quantity of micro-organisms ingested (μg C/10 cm^3/h)	
	bacteria	diatoms	bacteria	diatoms
Copepods	0.2170	0.477	0.5000	1.48
Nematodes	0.1400	2.290	0.3230	7.11
Ostracods	0.0277	0.870	0.0638	2.70
Polychaetes	33.0000	3.850	76.0000	11.90
Total	33.3847	7.487	76.8868	23.19

Data from Montagna (1984).

the model of Daro (1978) as verified by Roman and Rublee (1981) to determine grazing rates. Where,

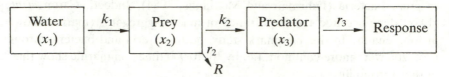

Here, k_1 and k_2 are transfer rates in hours, and represent the portion of radioactivity removed from the donor compartment per unit time. The flow rate of radioactive tracer, through the system, is given by:

$$\frac{dx_1}{dt} = -k_1 x_1$$

$$\frac{dx_2}{dt} = k_1 x_1 - k_2 x_2$$

$$\frac{dx_3}{dt} = k_2 x_2 - r_3 x_3$$

An assumption is made that the amount of radio-label in water, i.e. x_1, is not limiting. Daro (1978) showed that the grazing rate (k_2) is:

$$k_2 = \frac{2x_3}{x_2 t}$$

Using the Daro (1978) model, the flow of tracer by bacteria is:

$$\frac{dx_2}{dt} = k_1 x_1, -k_2 x_2 - r_2 x_2$$

Modelling predator–prey interactions is scarcely a new occurrence, insofar as the Lotka–Volterra predator–prey equations date to the 1920s (Lotka, 1925; Volterra, 1926). Williams (1980) used these equations:

$$-\frac{dN_1}{dt} = r_1 N_1 - p_1 N_1 N_2$$

$$-\frac{dN_2}{dt} = p_2 N_1 N_2 - d_2 N_2$$

where, N_1 = density of prey, N_2 = density of the predator, r_1 = prey intrinsic rate of increase, d_2 = predator mortality rate, p_1, p_2 = predation interaction constants.

5.4 Degradation of complex molecules

It is a fascinating and remarkable achievement that there is widespread ability among marine micro-organisms to attack and degrade (and, therefore, to remove from the natural environment) complex molecules of biotic and abiotic origin. This ability has obvious benefits in the elimination of pollutants, a topic which will be discussed further in Chapter 8. For the present, it is relevant to consider the diverse range of enzymic activity attributable to marine micro-organisms, and, in particular, to the bacteria. This ability includes the production of amylases, deoxyribonucleases, lipases and proteases. The overwhelming majority of aerobic marine bacteria appear to be capable of producing these enzymes.

Nucleic acids, the basal material of all living organisms, are widely distributed in seawater and marine sediments (see Maeda and Taga, 1973). This is matched by the presence of large numbers of DN'ase-producing bacteria. Maeda and Taga (1973) found up to 4.1×10^5 DN'ase-producing bacteria/g dry weight of sediment in Tokyo Bay. This number represented between 14 and 76% of the total population of aerobic heterotrophs. Similar proportions were also found in the Pacific Ocean and Sea of Japan (Maeda and Taga, 1974). However, up to 95% of the aerobic heterotrophic bacteria found in UK coastal water and sediment were attributed with DN'ase activity (Austin, unpublished data).

Kjelleberg and Håkansson (1977) examined the distribution of amylase-, lipase- and protease-producing bacteria in surface and sub-surface water off the Swedish coast, noting a greater proportion of the latter than the former. Between 43 and 100% of the aerobic heterotrophs produced lipase, with a similar proportion producing protease (47–100%). Lower proportions (17–88%) were amylolytic. Similarly high proportions of bacteria from marine sediments collected from the Atlantic seaboard of the USA degraded lipids and proteins (Nitkowski *et al.*, 1977). Furthermore Sieburth (1971) reported that bacteria from the neuston demonstrated marked lipolytic activity.

Other complex molecules are attacked by marine micro-organisms. For example, Araki and Kitamikado (1978) reported the distribution of mannan-degrading bacteria in coastal water and sediment (mannan is a polysaccharide). In this study, the numbers of mannan-degraders in water and sediment were estimated to be $0–1.0 \times 10^3$/ml (0–25% of the total number of aerobic heterotrophs) and $0–4.0 \times 10^5$/g (0–100% of the total bacterial population), respectively. It would appear that such populations equate well with the numbers of organisms which are capable of degrading xylan.

Bacteria have also been attributed with the ability to utilise dissolved carbohydrate in offshore waters. Thus Burney (1986) reported that total dissolved carbohydrate declined at 1.5 and 2.6 μg C/l/h in 12–24 hours during dark *in situ* incubations in the Sargasso Sea and Gulf of Maine, respectively. This decline was interpreted as indicating bacterial metabolism.

5.5   Colonisation of surfaces, and the role of chemotaxis and attachment

In this author's opinion, one of the most fascinating aspects of marine organisms is their ability to rapidly colonise barren surfaces. The implications of this phenomenon for biofouling will be discussed separately in Chapter 8. However, the ability is not newly recognised insofar as Zobell and Allen (1935) published a landmark article over half a century ago. For recent literature the reader should consult Fletcher (1980), Bright and Fletcher (1983), Hermansson and Marshall (1985) and Kjelleberg *et al*. (1985). Since 1935 scientists have devised elaborate procedures to investigate attachment. Yet there has been a fascination with glass surfaces, rather than other substrates, e.g. wood, which abound in the marine environment. Whatever the surface the scenario will inevitably involve an initial attraction (chemotaxis) to a site, followed by attachment and thence colonisation, the latter of which will be discussed in Chapter 8.

5.5.1   *Chemotaxis*

It is easy to visualise a number of important roles for chemotaxis, such as in predator–prey relationships (Walsh and Mitchell, 1978), as well as in colonisation of surfaces. Thus it may be assumed that chemotactic responses will be demonstrated by motile rather than sessile organisms to some form of attractant, possibly nutritional in origin. Here an inference is made that the organism in question will be attracted to possible sources of food. In carrying out experiments with a marine pseudomonad, Chet and Mitchell (1976) deduced that the attractants included low levels of *D*, *L* and *DL* forms of amino acids (it is relevant to note that many aerobic heterotrophic bacteria prefer to utilise amino acids rather than sugars). Indeed the threshold level for chemotaxis was 10^{-8} M in the case of leucine and cysteine. Thus in seawater, with dissolved organic material normally in the pico- to nanomolar range (Williams, 1986), there are obvious advantages in moving to areas of more favourable nutrient levels. Such habitats may include faecal pellets, which become readily colonised by bacteria (although they may be of intestinal origin) (Jacobsen and Azam, 1985), polysaccharide-rich areas, e.g. seaweeds, which are a major source of bacterial nutrition (Hagström *et al*., 1984), and especially the zone around algal cells. Azam and Ammerman (1984) proposed that micro-algae are surrounded by zones of high organic concentrations which are attributable to the leaky nature of the cells. In fact, algae leak as much as 5–30% of their photosynthates (Mague *et al.*, 1980). Similarly, microflagellates release amino acids and inorganic nutrients (Andersson *et al.*, 1985). Therefore it may be expected that bacteria migrate to such areas of heightened nutrient levels by chemotaxis.

It must be emphasised that the nature of chemotaxis is improperly understood. The phenomenon is clearly worthy of more extensive study to clarify some of the uncertainties enshrouding what is an ecologically important topic. Is there a relationship with the fascinating magnetotactic bacteria? Such organisms, which possess magnetite (Fe_3O_4) particles (these appear as lattice structures with hexagonal prisms; Mann *et al.*, 1984), are found in marine sediments. Magnetotactic bacteria orient and swim along geomagnetic fields, and possess a direction of magnetic polarity such that the organisms swim exclusively downwards in both the Northern and Southern Hemispheres (Mann *et al.*, 1984).

5.5.2 *Attachment*

The initial reaction between microbial cells and solid surfaces is the result of physico-chemical forces, of which the most important are Van der Waals forces of attraction, and electrostatic repulsive forces (Marshall, 1976). The factors affecting the initial interaction include the nature and electrical charge of the solid surface, and the surface components and charge of the microbial cells. Attachment occurs to a wide variety of surfaces (Fletcher and Loeb, 1979), although the precise mechanism is often unclear. There may be specialised attachment structures, such as fimbriae (pili) on some Gram-negative bacteria, e.g. *Vibrio*, and stalks/holdfasts/prosthecae (these occur on *Hyphomonas*, *Hyphomicrobium* and *Caulobacter*), or extracellular secretions (polymers). Numerous authors have implicated a role for extracellular polysaccharides (e.g. Corpe, 1970; Marshall *et al.*, 1971a, b; Fletcher and Floodgate, 1973), whereas others have investigated proteinaceous compounds (Fletcher, 1976). In the latter study, Fletcher (1976) investigated the role of proteins on the attachment of a marine *Pseudomonas*.

The importance of polymers in attachment has been extensively studied. Fletcher and Floodgate (1973) pointed to an initial involvement of discrete amounts of specific, albeit unnamed, polymers. Thence afterwards, i.e. after the initial contact and attachment has been achieved, a more extensive development of secondary polymers occurs (Corpe, 1970; Fletcher and Floodgate, 1973). Indeed, attachment fibres/fimbrils may be readily observed by scanning electron microscopy (Paerl, 1975; Dempsey, 1981). However, it should not be overlooked that such fibrils may be artefacts, i.e. strands of condensed polymer, which result during dehydration of the samples prior to observation by scanning electron microscopy. Nevertheless, Dempsey (1981) demonstrated the presence of condensed polymers which formed as a sheet or film overlying the fibrils. This observation substantiates the importance of fibrils in attachment by providing evidence that their presence is not reflectory of harsh dehydration processes.

It should also be considered that bacterial polymers, i.e. mucus, may function in the physical aggregation of sediment particles. In particular, bacterial mucus may enhance clay accumulation in sediments; this phenomenon appears to be influenced by temperature (De Flaun and Mayer, 1983).

5.6   Biogeochemical processes (nutrient cycling)

There is a perpetual need for compounds to be modified in order for them to be used by various classes of organisms. Organic material (carbon) will be mineralised, yielding products which are essential for dissimilatory and assimilatory processes. In other words, the existence of life in the marine environment is dependant upon the continual release of fresh supplies of essential nutrients. In pelagic waters recycling of minerals is a slow process, but greater rates of activity occur in sediments and in coastal waters. According to Jørgensen (1980) the cycling of elements is regulated by two main processes, namely the assimilation of inorganic nutrients by photosynthetic organisms, and the subsequent mineralisation by heterotrophs. Greatest attention has focused on the cycling of nitrogen, sulphur, carbon and phosphorus. The role of these elements in life-giving processes is well established. Thus, nitrogen is necessary for amino acids, nucleic acids and amino sugars; sulphur is essential in the sulphydryl groups (–SH) of amino acids and their polymers; carbon is the prime substance of all organic compounds; and phosphorus is contained in nucleic acids, phosphate esters, sugar phosphates, phosphates and ATP. These elements are taken up by autotrophs and phototrophs as mineral salts. Conventionally, nutrient cycling is described in terms of each element, a procedure which will be followed here.

5.6.1   *The nitrogen cycle*

A schematic representation of the nitrogen cycle has been included in Fig. 5.1. Nitrogen enters the cycle as ammonia, resulting from the breakdown of proteins by a process called ammonification, and as nitrates, which are derived from fertilisers etc., and arrive in the sea via terrestrial run-off. For example, for the UK the annual discharge of nitrates from estuaries into the sea has been calculated as 2×10^5 tonnes of nitrogen (Royal Society, 1983). In the southern Scotia Sea during 1981 there were high concentrations of nitrate (21–29 μM) in the water, although the resident phytoplankton utilised ammonium ions as the source of nitrogen (Koike *et al.*, 1986). In addition, there will be an input of fixed nitrogen amounting to 4×10^7 tonnes N/annum resulting from the activity of nitrogen-fixing organisms, notably cyanobacteria (Stal *et al.*, 1984) of the genera *Anabaena* and *Nostoc*. Some of the organisms possess heterocysts, which separate the oxygen-sensitive nitrogenase from the oxygen-producing photosynthetic system. Nitrogen fixation requires a considerable input of energy, i.e. 150 kcal/mol, as ATP and reduced co-enzymes (Bull, 1980). This energy is derived from respiration by heterotrophs, or from the conversion of solar energy in the case of photoautotrophs and cyanobacteria. Oceanic coccoid cyanobacteria possess large quantities of phycoerythrin, which presumably serve as nitrogen reserves as well as accessory pigment in photosynthesis (Fogg, 1986).

With ammonification there is the generation of ammonia from organic matter/proteins. An example follows (after Atlas and Bartha, 1987):

$$\text{urease}$$
$$H_2O + NH_2.CO.NH_2 \rightarrow 2NH_3 + CO_2$$

Heterotrophic micro-organisms generate ammonium ions which are used by the phytoplankton (Koike *et al.*, 1986). The ammonia/ammonium ions

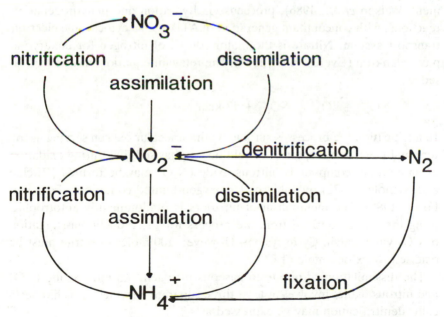

Fig. 5.1 A simplified diagram representing the nitrogen cycle (after Brown, 1987).

so formed may be oxidised (the process is known as nitrification) or assimilated directly into amino acids, either by glutamine synthetase/glutamate synthase or by direct amination of a α-keto-carboxylic acid. Approximately 78% of the total uptake of ammonium ions (NH_4^+) is by prokaryotes, notably heterotrophic bacteria. Conversely, regeneration of NH_4^+ is largely by eukaryotes (Wheeler and Kirchman, 1986).

The first stage in the oxidation of ammonia/ammonium ions (by nitrification), i.e. by representatives of *Nitrosococcus, Nitrosolobus, Nitrosomonas, Nitrosospira* and *Nitrosovibrio*, is a cytochrome-dependant, energy-requiring hydroxylation to form hydroxylamine (NH_2OH) and possibly other as yet unknown intermediates. Further oxidation of hydroxylamine to nitrite is coupled to the generation of ATP, via a membrane-bound electron transport system, which contains cytochromes and flavins. Overall the reaction may be expressed as:

$$NH_4^+ + 1\tfrac{1}{2}O_2 \rightarrow NO_2^- + 2H^+ + H_2O + 66 \text{ kcal}$$

Evidence from experiments with batch cultures suggest that methane inhibits the oxidation of ammonia, in the case of *Nitrosococcus oceanus* (Ward, 1987). Oxidation of nitrite to nitrate, which is achieved by representatives of *Nitrobacter, Nitrococcus, Nitrospina* and *Nitrospira* (*Nitrospira* is possibly one of the most common nitrite oxidisers in the oceanic environment; Watson *et al.*, 1986), proceeds via hydration and dehydrogenation reactions, with concomitant generation of ATP from a cytochrome electron transport system. Nitrate is the major source of nitrogen for eukaryotic phytoplankton (Everest *et al.*, 1984). The relevant equation may be expressed as:

$$NO_2^- + \tfrac{1}{2}O_2 \rightarrow NO_3^- + 17 \text{ kcal}$$

In nitrification the process is aerobic, with molecular oxygen serving as the terminal electron acceptor. It should be emphasised that during oxidation of ammonium compounds, nitrous oxide (N_2O) may be formed (Ritchie and Nicholas, 1972), possibly under oxygen-limited conditions (Poth and Focht, 1985). The metabolism of nitrifiers is predominantly autotrophic, using the energy derived from the process for the reduction/assimilation of CO_2 via a Calvin Cycle system. However, 100 moles of nitrite must be oxidised to fix one mole of CO_2.

The dissimilation of nitrate to gaseous nitrogen, via nitric oxide (NO) and nitrous oxide, involves at least three separate enzymes. Stoichiometrically, denitrification may be expressed as:

$$5C_6H_{12}O_6 + 24NO_3^- + 24H^+ \rightarrow 30CO_2 + 12N_2 + 42H_2O$$

Organisms which carry out the reaction include the majority of marine aerobic and facultatively anaerobic heterotrophic bacteria together with many algae and fungi. The dissimilation to nitrite may also involve sulphate-reducing bacteria. In marine sediments there is a constant accumulation of ferrous ions (Fe^{2+}) (following the reduction of ferric (Fe^{3+}) compounds), immediately after the depletion of endogenous nitrates (Sørensen, 1982). It is considered likely that the process occurs as a result of activity by facultatively anaerobic nitrate reducers.

Marine picoplankton, namely cyanobacteria, are capable of utilising nitrate, ammonium compounds and/or urea as the sole source of nitrogen for growth (Probyn, 1985; Probyn and Painting, 1985). In particular picoplankton are responsible for the uptake of the majority (90%) of nitrogen accounted for by phytoplankton in the oceanic environment (Stockner and Antia, 1986). It appears that the major pathway of inorganic nitrogen assimilation by chroococcoid cyanobacteria is nitrate to nitrite, and thence to ammonium, glutamine and finally to glutamic acid (Syrett, 1981).

5.6.2   *The sulphur cycle*

The sulphur cycle has been summarised in Fig. 5.2. Sulphate, which is assimilated by many micro-organisms, is the second most abundant anion in seawater. However, to be incorporated into macro-molecules, i.e. cys-

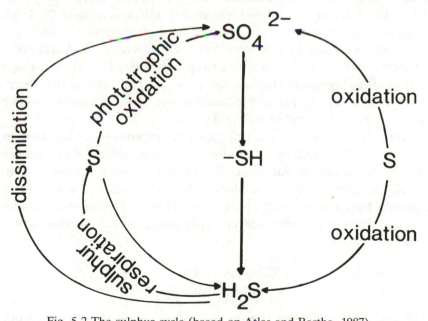

Fig. 5.2 The sulphur cycle (based on Atlas and Bartha, 1987).

teine, methionine and co-enzymes, sulphate needs to be reduced to sulphide. Sulphur may be released as mercaptans and hydrogen sulphide during the decomposition of organic matter, for example:

cysteine
$$HSCH_2.CHNH_2.COOH + H_2O \rightarrow HOCH_2.CHNH_2.COOH + H_2S$$
desulphydrase

In the presence of oxygen, these reduced sulphur compounds, e.g. sulphides, are capable of supporting chemolithotrophic metabolism. Thus in the case of bacteria, such as *Beggiatoa, Thiothrix* and *Thiovulum*, sulphur droplets are deposited intracellularly following the reaction:

$$\dot{H}_2S + \tfrac{1}{2}O_2 \rightarrow S + H_2O + 50.1 \text{ kcal/mol}$$

In environments at the interface of aerobiosis and anaerobiosis these droplets may be oxidised to sulphate.

In the case of thiobacilli, most of which are aerobic, some species oxidise hydrogen sulphide and other reduced sulphur compounds with the production of sulphur droplets; whereas other taxa produce sulphate from the oxidation of elemental sulphur as follows:

$$S + 1\tfrac{1}{2}O_2 + H_2O \rightarrow H_2SO_4 + 149.8 \text{ kcal/mol}$$

Such thiobacilli are obligate chemolithotrophs, obtaining energy solely from the oxidation of inorganic sulphur compounds, and carbon from CO_2. From standard values of growth yield, a maximum carbon dioxide assimilation of 5 mMol C/m^2/day in sediments has been calculated for Limfjorden, Denmark; an amount corresponding to approximately 15% of the average daily input of organic matter into the sediment (Fenchel and Blackburn, 1979). Indeed, the energy and reducing power of the chemolithotrophic activity is used to assimilate carbon dioxide.

Hydrogen sulphide is also subjected to phototrophic oxidation in anaerobic environments by photosynthetic sulphur bacteria of the families Chromatiaceae and Chlorobiaceae. These organisms occur in sulphide-containing environments where sulphate-reducers are active. They occur in especially large numbers in the coastal environment. The organisms photoreduce carbon dioxide while oxidising hydrogen sulphide to sulphur according to the reaction:

$$nCO_2 + nH_2S \rightarrow (CH_2O)n + nS$$
photosynthate

In anaerobic sediments rich in sulphide, organisms such as *Desul-*

furomonas grow on acetate and anaerobically reduce sulphur to hydrogen sulphide, according to the equation:

$$CH_3COOH + 2H_2O + 4S \rightarrow 2CO_2 + 4H_2S + 5.7 \text{ kcal/mol}$$

Desulfuromonas co-exists also with phototrophic green sulphur bacteria that photo-oxidise hydrogen sulphide to elemental sulphur, which is excreted externally to the cells.

Sulphate-reducers, which carry out the reaction,

$$4H_2 + SO_4^{2-} \rightarrow H_2S + 2H_2O + 2OH^-$$

(Stoichiometrically, sulphate reduction may be expressed as:

$$C_6H_{12}O_6 + 3SO_4^{2-} + 6H^+ \rightarrow 6CO_2 + 3H_2S + 6H_2O)$$

include representatives of *Desulfobacter, Desulfobulbus, Desulfococcus, Desulfosarcina* and *Desulfovibrio*. The reaction is strictly anaerobic, and inhibition occurs in the presence of oxygen, nitrate and ferric compounds. Sulphate reduction, which may be plasmid-mediated (Postgate, 1984), occurs predominantly in sediments, particularly in the upper 20 cm layer where methanogens are outcompeted (Nedwell, 1984). However, the importance of sulphate reduction declines with increasing depth into the sediment, and increasing distances from shore. Nevertheless, the activity of sulphate-reducing bacteria (SRB) should not be underrated, insofar as the organisms remove much of the approximately 10^{11} kg of sulphate which enters the marine environment each year from rivers as a result of terrestrial erosion. Sulphate reduction and Ca^{2+} precipitates serve to maintain the concentration of sulphate in seawater at 29 nmol. This is reflected in the constant ratio of $SO_4:Cl^-$ at 0.14 $\pm$0.4% (Wilson, 1975).

In sediments, respiration of sulphate is the dominant form of anaerobic oxidation, accounting for approximately half of the mineralisation of total organic carbon (Jørgensen, 1977; Nedwell, 1984). In fact within coastal sediment SRB may oxidise as much organic matter to carbon dioxide as the process of aerobic respiration (Jørgensen, 1982). The substrates (electron donors) used for the dissimilatory reduction of sulphate in sediments are principally the fermentation end-products (notably molecular hydrogen and short chain fatty acids) of other organisms. The relative importance of each electron donor has been the focus of study by Sørensen *et al.* (1981) and Christensen (1984). These workers suggested that the contribution of the various substrates to sulphate reduction – on a percentage basis – was:

acetate	40–65%
butyrate	5–20%
hydrogen	5–10%

isobutyrate   8%
propionate   10–20%

Clearly the importance of acetate is highlighted, particularly if oxidation of butyrate and propionate is incomplete. In this case, the oxidation of acetate will account for approximately two-thirds of the sulphate reduction occurring in sediments. Essentially, the role of SRB in anaerobic food chains may be summarised as the oxidation of acetate and other substrates, and the utilisation and, indeed, production of hydrogen (the latter phenomenon occurs under conditions of sulphate limitation) (Peck and Odom, 1984). Incidentally, it should be emphasised that evidence has been published concerning the oxidation of other small molecules, notably ethanol, lactate, malate, pyruvate and succinate (Jørgensen, 1980). However the significance of these compounds to SRB is unclear. Furthermore SRB do not oxidise carbohydrates and hydrocarbons, because of the absence of a complete TCA cycle.

SRB are the principal methylators of Hg^{2+} in anoxic estuarine sediments. Here, methylation leads to the production of toxic methylmercury compounds (Compeau and Bartha, 1985). The formation of such compounds is one of the obvious malefits of marine bacteria, the significance of which will be discussed separately in Chapter 8.

5.6.3   *Carbon cycling*

In seawater there are 34 500 billion tonnes of carbon, the cycling of which is in a steady state (Hobbie and Melillo, 1984). Carbon transfer through food webs is summarised in Fig. 5.3; whereas carbon flow through anaerobic sediments is represented in Fig. 5.4. Carbon dioxide fixation to

Trophic level

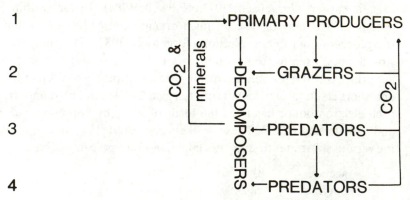

Fig. 5.3 Carbon transfer through a food web (after Atlas and Bartha, 1987).

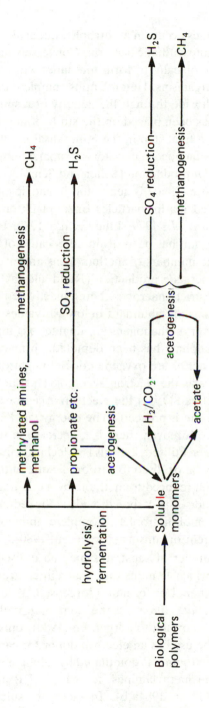

Fig. 5.4 Carbon flow through an anaerobic marine sediment ecosystem (after Parkes, 1987).

form organic molecules occurs in autotrophs, i.e. algae, cyanobacteria and green and purple photosynthetic bacteria. Conversely methanogens reduce carbon dioxide anaerobically to form methane, which is in turn used by comparatively few organisms. Heterotrophs complete the carbon cycle, by generating carbon dioxide through the activity of respiration.

Great emphasis has been placed on the study of methanogenesis, which may be regarded as a final step in the mineralisation of organic matter. It has been recorded that oceanic water is supersaturated with methane, attributed to its *in situ* production (Seiler and Schmidt, 1974). Suggestions have been made that methane is generated in anoxic micro-environments, provided that so-called 'marine particles' or zooplankton are present (Oremland, 1979; Traganza *et al.*, 1979; Lilley *et al.*, 1982; Burke *et al.*, 1983). However this suggestion has been disputed because of calculations which demonstrate that the numbers of methanogens are insufficient to explain the level of *in situ* methane production (Rudd and Taylor, 1980).

Methanogens are strict anaerobes, being inactivated by the presence of oxygen. The organisms are abundant below Eh values of -200 mV (Mah *et al.*, 1977). Moreover methanogens are unable to compete with sulphate-reducers until the sulphate has been depleted. In marine sediments only the upper few millimetres are oxygenated. Below this layer for a few centimetres, NO_3^- serves as the electron acceptor. Then below this level for a further few centimetres SO_4^{2-} is the electron acceptor and, in the deeper sediment, methanogenesis proceeds. The energy yield also decreases from oxygen to methane. Moreover, there is a decrease in the range of energy substrates which can be utilised, i.e. this is a decline in nutritional versatility. Organic carbon may be oxidised to carbon dioxide in the anaerobic layers of sediment, and then reduced to methane by methanogens. This methane is undoubtedly re-oxidised to carbon dioxide at the sediment surface, before assimilation by chemolithotrophs. Complete mineralisation inevitably involves many different organisms (Jørgensen, 1980).

Only a limited number of substrates may be utilised by methanogens. There are reports which demonstrate that either acetate (Sansone and Martens, 1981), methanol or formate (Jørgensen, 1980), or hydrogen/carbon dioxide (Lein *et al.*, 1981) are the dominant methane precursors in marine sediments. According to Jørgensen (1980), only hydrogen is sufficiently reducing to be used as an electron donor for carbon dioxide reduction. However, in environments dominated by sulphate reduction, methane may be derived from methylamines, for which SRB are not competitive (King *et al.*, 1983; King, 1984a,b). In fact both sulphate-reducers and methanogens are dependant on substrates which are the products of fermentation reactions in the sediment. Both groups of organisms will be competing for acetate and hydrogen (Jørgensen, 1980).

Methylotrophs are also present in marine sediments. These organisms couple methane oxidation with reduction of sulphate or elemental sulphur (Hanson, 1980; Large, 1983). A novel marine methylotrophic methanogen, named *Methanococcus frisius*, has been recovered from mud in the southern North Sea. This organism grew on hydrogen, carbon dioxide, methanol and methylamines (Blotevogel *et al.*, 1986).

5.6.4 *Other cycles*

Beside the extensively studied albeit incompletely understood carbon, nitrogen and sulphur cycles, information has been published about the cycling of iron, manganese and phosphorus. Certainly it is well established that some bacteria oxidise iron:

$$Fe^{2+} \rightarrow Fe^{3+} + 11.5 \text{ kcal/mol}$$

and use the energy to reduce carbon dioxide. Other organisms, e.g. *Pedomicrobium*-like bacteria, precipitate iron or manganese as sheaths (Ghiorse and Hirsch, 1982). Sørensen (1982) correlated the reduction of ferric to ferrous ions in sediments with the activity of nitrate-reducers.

The importance of phosphorus to biological systems is illustrated by its presence in nucleic acids. Phosphorus is a limiting nutrient in marine habitats, although the turnover time is often very short (Campbell, 1981). Many micro-organisms have the ability to degrade organic phosphorus compounds, with the release of phosphates. Indeed, these are intracellular storage products for some bacteria. However it should not be overlooked that one of the principal sources of phosphorus is bones, of which a key ingredient, calcium phosphate $(Ca_3(PO_4)_2)$, is attacked by some organisms.

Finally, it should be emphasised that the aforementioned nutrient cycles do not operate in isolation from each other. Instead there is a pronounced interaction, which leads to the continual cycling of essential elements and, thus, the continuance of life as we know it.

5.7 Primary productivity

Every student of biology will be aware of the role of photosynthesis in the production of organic matter by plants as a result of the green photosynthetic pigments, i.e. chlorophyll, and the interaction with carbon dioxide and water. Simple equations adorn basic biology textbooks, demonstrating that, with CO_2 and H_2O, the end result is carbohydrate. Thus, the role of green terrestrial plants as the saviours of animal life on

Earth would seem undisputed. Yet of the total estimated plant primary productivity, i.e. 1.4×10^{14} kg dry wt/year, 40% is derived from the activity of marine phytoplankton, of which the role of algal picoplankton (size range of 0.2–2.0 µm) is of overwhelming importance (Sherr and Sherr, 1984; Stockner and Antia, 1986). In fact, the hourly carbon production rates for marine picoplankton have been calculated to be in the range of 0.0004 mg C/m^3 (Saijo and Takesue, 1965) to >31 mg C/m^3 (Glover *et al.*, 1985b) (Table 5.2). These quantities correspond to 0.1–90% of the total carbon production in the marine environment, with higher values recorded for oligotrophic areas of the open ocean. Moreover primary production has been noted to increase with depth into the euphotic zone (Li *et al.*, 1983; Platt *et al.*, 1983; Glover *et al.*, 1985a,b), which partially reflects a greater efficiency of the photosynthetic pigments to utilise blue-green light. Certainly there are daily and seasonal fluctuations in rates of primary productivity. For example, Glover *et al.* (1985b) recorded higher carbon assimilation rates for algae in surface waters in the periods just prior to dawn and dusk. Moreover, on bright days, there was apparent photo-inhibition during the period close to midday. This information suggested to Glover and co-workers that there was greater efficiency of photosynthesis during periods of low photon flux. Conversely for cyanobacteria, Putt and Prezelin (1985) recorded peak rates of productivity at midday. Certainly the seasonal pattern of primary production was highlighted by Joint *et al.* (1986), who recorded peak levels in temperate waters during summer, i.e. August.

According to data published byWoodwell (1970), net primary productivity for oceanic, coastal and upwelled seawater is 100, 200 and 600 g dry wt of organic material/m^2/year, respectively. Here, the relative importance of upwelling areas in marine primary productivity is clearly demonstrated.

Overall it is considered that algae, i.e. diatoms, dominate by contributing 20–25% of the world's net primary productivity. However the importance of cyanobacteria should not be overlooked, with one genus, *Synechococcus*, responsible for 10% of the total marine primary productivity (Waterbury *et al.*, 1979). These members of the phytoplankton community are especially abundant in the neuston. Thus CO_2 fixation (resulting from mineralisation of organic material or from the atmosphere) with the production of organic molecules results from the activity of autotrophs, namely algae (diatoms) and cyanobacteria, as mentioned above and also from the activity of green and purple photosynthetic bacteria. Even the nitrite oxidiser, *Nitrospira*, has been credited with the ability to fix CO_2 (Watson *et al.*, 1986). Here the principal metabolic pathway involves the pentose phosphate or Calvin Cycle. However, photosynthetic organisms are also capable of utilising CO_2

Table 5.2 *Productivity of marine algal picoplankton*

Sampling site	Fraction size (μm)	Production rate (mgC/m³/h)	% of Total production	References
Atlantic Ocean				
Nova Scotia Shelf	0.2–1.0	1.8–5.8	12–65	Douglas (1984)
Arctic	0.2–1.0	0.12–2.0	10–25	Smith *et al.* (1985)
North West	0.2–1.0	0.0004–0.65	0.1–66	Glover *et al.* (1986)
Sargasso Sea	0.2–3.0	3.1–31.0	60–80	Glover *et al.* (1985b)
Equatorial	0.45–3.0	1.08	20	Herbland and le Bouteiller (1981)
Tropical	0.4–1.0	0.5	50–60	Platt *et al.* (1983)
Indian Ocean				
South China Sea	0.45–0.8	0.01–0.08	1–22	Saijo and Takesue (1965)
South China Sea	0.8–5.0	0.02–0.16	9–40	Saijo and Takesue (1965)
Pacific Ocean				
Tropical	0.2–1.0	0.2–45	20–80	Li *et al.* (1983)
Subtropical	0.2–3.0	0.75–1.5	70	Bienfang and Takahashi (1983); Takahashi and Bienfang (1983)
Subtropical	0.45–5.0	7.6	83	Bienfang and Szyper (1981)
Coastal California	0.2–5.0	1.29–2.19	75	Putt and Prezelin (1985)
North Central Gyre	0.45–2.0	1.86	57	Iturriaga and Mitchell (1986)
Baltic Sea	0.2–3.0	0.2–3.4	3–43	Larsson and Hagström (1982)
Oslofjorden (Norway)	0.45–5.0	0.1–30	15–90	Throndsen (1978)
Celtic Sea	0.2–2.0	0.7–12.4	20–30	Joint and Pomroy (1983)
Mediterranean Sea	0.2–3.0	2.5–9.2	25–90	Berman *et al.* (1984)

After Stockner and Antia (1986).

through the phosphoenol pyruvate carboxylase system. Nevertheless not all this primary production is available to marine food chains/webs, insofar as some organic matter is converted back to CO_2 during respiration.

Bacterial productivity measurements have been the focus of intense interest. Turley and Lochte (1985) observed that, during summer in the Irish Sea, bacterial productivity was greater in the waters above the thermocline than below it. Thus at 4 and 60 m depth, bacterial productivity was determined to be 12.7 µg C/l/day (generation time = 0.9 day) and 3.0 µg C/l/day (generation time = 20 days), respectively. Using the incorporation of tritiated thymidine into DNA as a criterion, Moriarty *et al.* (1985) estimated bacterial production as 120–370 mg C/m^2/day during summer in water over a coral reef flat. This amount equated to 30–40% of the primary productivity by benthic microalgae. Interestingly during winter, productivity was only 20% of the value for summer. Lower values were published by Alldredge *et al.* (1986), who recorded total bacterial productivity amounting to only 3–20% of the productivity in surface water. Within sediments bacterial productivity (as measured by the rate of tritiated thymidine incorporation into DNA, and the rate of ^{32}P incorporation into phospholipid) was estimated as 180–190 mg C/m^2/day (Moriarty *et al.*, 1986). Significantly most bacterial production occurred in the top 20 mm layer of sediment, which was the zone of greatest root and rhizome biomass. In the site which was the focus of study, seagrass (*Halodule wrightii*) was present. Seagrasses are considered to be major primary producers in shallow coastal temperate and tropical regions. Moreover in this study, 6–17% of fixed carbon was exuded from seagrass into the sediment. Thus the high value of bacterial productivity was attributed to exudation as well as root decomposition (Moriarty *et al.*, 1986).

Generally, it would appear that habitats which are capable of retaining nutrients, e.g coral reefs, are able to maintain high levels of productivity, even if the surrounding environment is oligotrophic, i.e. nutrient-limited/ poor. Conversely, systems with low capacities to retain essential nutrients, e.g. epipelagic habitats, demonstrate low nutrient-limited rates of primary productivity, even if light and temperature conditions favour high productivity (Odum, 1983).

6   Microbiology of macro-organisms

It may be reasoned that the numbers and types of micro-organisms considered to be in and around healthy and diseased plants and animals will reflect the nature of the methods employed. This point is of paramount importance in any consideration of the microflora of living organisms. Macro-organisms are continually bathed in an aqueous suspension of microbes. Therefore, external surfaces will have frequent contact with these micro-organisms. It is apparent that micro-organisms may:

(a) become intimately associated with the external surface(s) of the plant or animal. Here, the potential exists for the micro-organisms, to colonise the surface and beome part of a hypothetical microflora. This microflora could inhibit or retard colonisation by other micro-organisms.

(b) be inhibited by any anti-microbial compounds which are produced by the prospective host. For example, recent evidence suggests that fish mucus contains anti-bacterial compounds (Austin and McIntosh, in press).

(c) accumulate at the site of any damage on the host. Such damaged areas could leak nutrients, which permit the growth of micro-organisms. These microbes may be subsequently incorporated as part of a resident microflora.

(d) enter the host through natural openings, e.g. the mouth. Here the micro-organisms may colonise the host, be inhibited or even digested, or pass again into the outside aquatic environment.

(e) pass by the macro-organism in the water column.

The micro-organism may be actively attracted to the host by chemotaxis, or make accidental contact. Any subsequent association will be either beneficial, neutral or harmful to the host.

Most of the data have been derived from simplistic experiments, usually involving microscopy and/or culturing techniques. However, the reliability of such data may be questionable. For example, the surface of the macro-organism may be swabbed, and the material used as inoculum for agar media. In this case it would be unclear if the resulting colonies were derived from organisms, which where closely adhered to or colonising the surface, were transient, or merely contained in the water film surrounding the macro-organism. Similarly, homogenates of internal structures, notably the intestines of marine animals, may reveal only food-borne micro-organisms rather than true inhabitants. Nevertheless using such techniques, considerable information has been amassed about the microflora of marine macro-organisms.

6.1   Microbiology of plants

Marine plants have received limited attention from microbiologists. Electron microscopy has revealed a dense microflora comprising diatoms, yeasts and filamentous bacteria on the surface of the red alga *Polysiphonia lanosa* (Sieburth *et al.*, 1974). This is an epiphyte on the brown alga *Ascophyllum nodosum*. Numerically, bacterial populations have been estimated to be in the range of 10^1–10^7/g dry weight. Thus, Chan and McManus (1969) reported 10^4–10^7 bacteria/g dry weight on the surface of *Ascophyllum nodosum* and *Polysiphonia lanosa* whereas Conover and Sieburth (1964) calculated population densities of 2×10^1–2×10^5 bacteria/g dry weight of *Sargassum natans*. Similarly, Kang (1981) recovered 1.1×10^4 bacteria/g from healthy fronds of the seaweed *Undaria pinnatifida*. Generally these counts were considered to be much higher than in the surrounding seawater. Perhaps a more meaningful estimate of population density equates numbers of micro-organisms per unit area of plant surface. Using this system, bacterial populations appear to be in the range of 10^1–10^6/cm^2 depending upon the season and the nature of the plant examined. Laycock (1974) determined the bacterial populations on *Laminaria* to be 10^1–10^3/cm^2. On brown alga (*Eisenia bicyclis*), green alga (*Monostroma nitidum* and *Enteromorpha linza*) and red alga (*Porphyra suborbiculata*) the bacterial numbers were 2.9×10^1–2.2×10^4/cm^2, 10^3–10^4/cm^2 and 10^3–10^4/cm^2, respectively (Shiba and Taga, 1978). However, the counts for green and red algae were constant during a seven month period, whereas there was a fluctuation in bacterial numbers attributed to the physiology of the host, on brown alga, with minimal and maximal populations in January and March, respectively. The smallest populations occurred when the largest number of germinating leaves was observed (Shiba and Taga, 1978). This supported the earlier observations of Sieburth *et al.* (1974), who noted that *Ascophyllum nodosum* possessed scant microbial populations during the growth period of spring and early summer, whereas in winter the microbial numbers increased. The reasons for this fluctuation may reflect changing nutrient levels on the plant surfaces or indicate the presence of inhibitory compounds during parts of the year. In this respect, polyphenols occur in *Ascophyllum nodosum* (Sieburth *et al.*, 1974) and tannic acid and tannin-like substances, which are inhibitory to bacteria, are present in *Sargassum natans* (Sieburth and Conover, 1965).

Evidence suggests that grasses possess a narrower range of micro-organisms than seaweeds. Thus, Sieburth *et al.* (1974) considered that cord grass, *Spartina alterniflora*, was colonised initially by fungi notably *Sphaerulina pedicellata*, whereas eel grass (*Zostera marina*) mostly posses-sed the pennate diatom, *Cocconeis scutellum*. On seaweeds, diatoms, yeasts and bacteria occur. Sieburth *et al.* (1974) reported the presence of micro-col-onies of *Leucothrix mucor* on *Ascophyllum nodosum* during winter. *Vibrio* spp. were considered to predominate on *Ascophyllum nodosum* (Chan and McManus, 1969) and *Laminaria longicruris* (Tsukidate, 1971). In contrast, *Aeromonas, Flavobacterium* and *Staphylococcus* were present on *Porphyra leucostica* (Tsukidate, 1971). According to Shiba and Taga (1978) yellow and orange-pigmented bacteria predominated on green and red algae. In the case of green algae, the dominant bacteria were representatives of *Cytophaga-Flavobacterium* with smaller numbers of *Acinetobacter*, coryneforms, *Pseudomonas* and *Vibrio*. Some of these bacteria may be beneficial to the host by supplying growth substances whereas other micro-organisms may be harmful, causing disease.

Kang (1981) discussed diseases of the seaweed, *Undaria pinnatifida*. Com-pared to healthy fronds, damaged areas (characterised by the presence of green spots) possessed 10 to 100-fold greater bacterial populations, i.e. $6.8 \times 10^5 - 1.2 \times 10^6/g$. These bacteria were equated with *Achromobacter* (a genus of dubious taxonomic validity), *Acinetobacter, Flavobacterium, Moraxella, Pseudomonas* and *Vibrio*. However the precise role of these taxa in the disease process is unclear. Nevertheless, Kang (1981) considered that pinhole-sized lesions in fronds were probably caused by a harpacticoid copepod, *Thalestris* sp.

6.2 Microbiology of healthy vertebrates

Using fish as the model, it is clear that the number and types of micro-organisms, notably bacteria, on the surface equate well with popula-tions in the surrounding water. Horsley (1973) reported populations of between 10^2 and 10^3 bacteria/cm^2 on the surface of Atlantic salmon (*Salmo salar*) caught in Britain. Higher values of between 4×10^3 and 8×10^4 bacteria/cm^2 were determined to be present on the surface of freshly caught flathead mullet (*Platycephalus fiscus*) and whiting (*Sillago ciliato*) in Australia (Gillespie and Macrae, 1975). However, in the case of turbot (*Scophthalmus maximus*), scanning electron microscopy has failed to support the presence of intimate colonisation of the surface (Figs. 6.1, 6.2) (Austin, 1982, 1983), although it must be accepted that preparatory procedures for microscopy may remove loosely associated micro-organisms from the surface of the specimens. Thus, the implication from electron microscopy is that most of the microbes are only loosely associated with the host's surface. The taxonomic composition of the surface microflora includes *Acinetobacter calcoaceticus, Alcaligenes faecalis, Bacillus cereus, B. firmus, Caulobacter,* coryneforms, *Cytophaga/Flexibacter, Escherichia coli, Hyphomicrobium vulgare, Lucibacterium harveyi, Photobacterium angustum, Ph. logei, Pseudomonas fluorescens, Ps. marina, Prosthecomicrobium and Vibrio* (Austin, 1982). The microbiology of fish muscle is improperly understood, insofar as it remains unclear whether or not the tissues are normally sterile. This aspect will be considered further in relation to fish spoilage (see Chapter 8).

Emphasis has been placed upon the microbiology of the gastro-intestinal tract of marine fish (Figs. 6.3–6.6). From the Japanese literature, it is

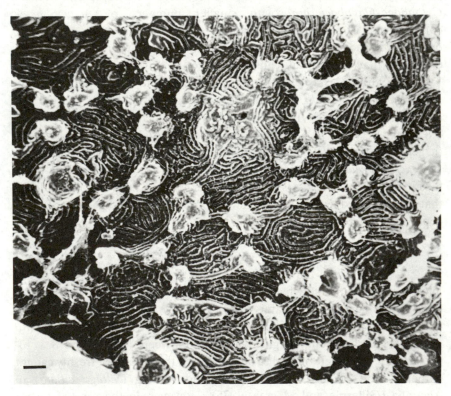

Fig. 6.1 Scanning electron micrograph of the surface of healthy turbot (*Scophthalmus maximus*). Bar equals 1 μm.

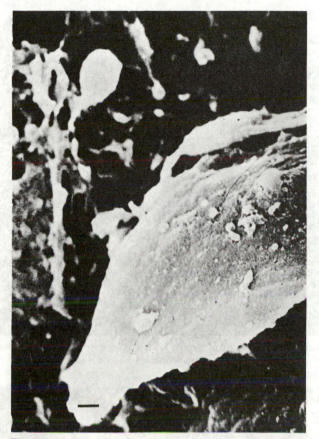

Fig. 6.2 Surface of turbot as revealed by scanning electron microscopy. Bar equals 1 μm.

apparent that microbial populations of up to 10^8 bacteria/g occur in the intestines. Such numbers appear to be much higher than in the surrounding water, indicating the presence of favourable ecological niches for microbes within the gastro-intestinal tract. There is some evidence for a seasonal fluctuation in numbers, with minimal and maximal numbers recorded in winter and summer, respectively. Differences have been noted in the microbial populations within separate regions of the gastro-intestinal tract of yellowtail (*Seriola quinquiradiata*). Concerning aerobic heterotrophic bacteria, the lowest counts occurred in the pyloric caeca (2×10^4 bacteria/g), intermediate numbers in the stomach (2.5×10^5 bacteria/g) and the highest populations generally in the intestines (6.5×10^4–5.9×10^6 bacteria/g) (Sakata *et al.*, 1978). This suggests that there is a continual movement of bacteria along the gut. The same workers recorded an additional 7.1×10^5 anaerobic bacteria/g from the intestines. Unfortunately, scepticism should

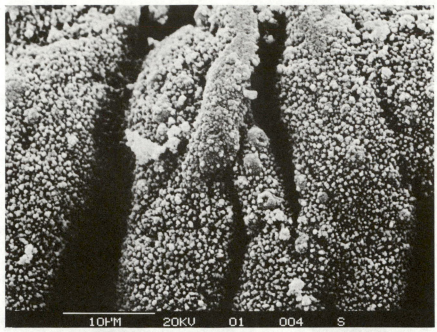

Fig. 6.3 Scanning electron micrograph of the gastro-intestinal tract of Dover sole. Bar equals 10 μm. Photograph courtesy of Dr. N.L. MacDonald.

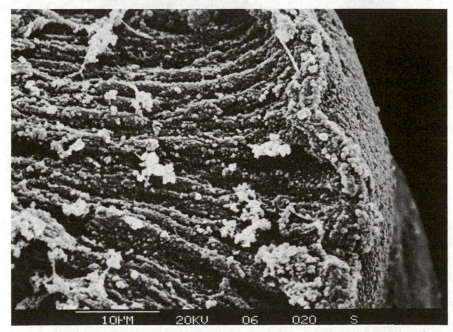

Fig. 6.4 Scanning electron micrograph of the gastro-intestinal tract of Dover sole. Bar equals 10 μm. Photograph courtesy of Dr. N.L. MacDonald.

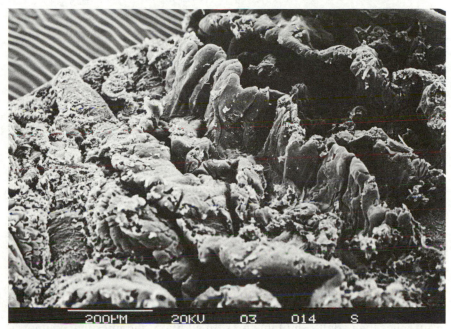

Fig. 6.5 Scanning electron micrograph of the gastro-intestinal tract of Dover sole. Bar equals 200 μm. Photograph courtesy of Dr. N.L. MacDonald.

Fig. 6.6 Scanning electron micrograph of the gastro-intestinal tract of Dover sole. Bar equals 20μm. Photograph courtesy of Dr. N.L. MacDonald.

be reserved for much of the methodology that has been used. For example, some workers have transported the fish or the entire gastro-intestinal tract in a frozen or fresh condition before examination, often many hours later. In such cases, the likely microbiological changes in the samples after capture and prior to examination have been ignored. Often the gut has been homogenised together with its contents, in which case the data may reflect micro-organisms in food, attached to the wall, or a combination of both of these possibilities. Noting such deficiencies in the published data, attention has been centred upon the microbiology of the gastro-intestinal tract of Dover sole (*Solea solea*) (MacDonald *et al.*, 1986). From freshly killed animals, the guts were withdrawn and divided into four sections; stomach, foregut, midgut and hindgut/rectum. Sections were opened lengthways, and gently agitated for 2–3 minutes in diluent to remove the contents. Gut sections and contents were then homogenised separately, and aliquots spread over the surface of marine 2216E agar plates. After aerobic incubation at 15 °C for 14 days, colony counts showed that there was a progressive increase in the numbers of bacteria along the gastro-intestinal tract, in agreement with the work of Sakata and colleagues, with higher populations in juveniles than adult fish (Table 6.1). In addition, the intestinal contents contained approximately 4.8×10^4 bacteria/g.

Taxonomic studies of the gastro-intestinal tract have revealed that there is generally a restricted range of micro-organisms, i.e. *Aeromonas, Pseudomonas* and *Vibrio* (Liston, 1957; Ugajin, 1979; Sakata *et al.*, 1980), which has led to suggestions that the gut flora is subjected to selective effects. However, MacDonald *et al.* (1986) reported a greater range of taxa in the gut of Dover sole, including *Acinetobacter, Alcaligenes*, Enterobacteriaceae representatives, *Flavobacterium, Micrococcus, Photobacterium, Staphylococcus* and *Vibrio*, with the latter predominating, i.e. 45% of the total number of isolates. Although the majority of these taxa were generally

Table 6.1 *Bacterial populations in farmed Dover sole*

Sample	No. of bacteria/g
Juvenile fish:	
Stomach/foregut	5.2×10^5
Midgut	8.0×10^5
Hindgut/rectum	9.8×10^6
Adult fish:	
Stomach	3.0×10^4
Foregut	3.0×10^4
Midgut	7.0×10^4
Hindgut/rectum	2.3×10^5

After MacDonald *et al.* (1986)

distributed throughout the gastrointestinal tract, *Alcaligenes* and *Staphylococcus* were not recovered from the hindgut and stomach, respectively. Fewer taxa (*Flavobacterium, Photobacterium* and *Vibrio*) were present in the gut contents, which indicates that some bacteria, i.e. *Acinetobacter, Alcaligenes*, Enterobacteriaceae representatives, *Micrococcus* and *Staphylococcus*, may be in intimate contact with the lining of the gastrointestinal tract. The possible role of these bacteria in the gut has received limited attention. Thus they may be regarded as beneficial to the host, in terms of nutrition by producing vitamins (Kashiwada *et al.*, 1971) and enzymes, e.g. chitinases, which contribute to the digestion of complex food particles (MacDonald *et al.*, 1986).

6.3 Microbiology of healthy invertebrates

There is conflicting evidence concerning the numbers, types and roles of micro-organisms on the surface of and inside the digestive tract of invertebrates. In part the dilemma appertains to whether or not data have been amassed from laboratory-maintained or wild-caught animals. Generally, it would appear that animals which have been maintained in laboratory facilities possess less micro-organisms than their wild-caught counterparts. For example, Austin and Allen (1982) ascertained that there were a dearth of micro-organisms on the surface of or within laboratory hatched *Artemia* nauplii (Figs. 6.7, 6.8). Indeed, the internal flora was estimated at 50 bacteria

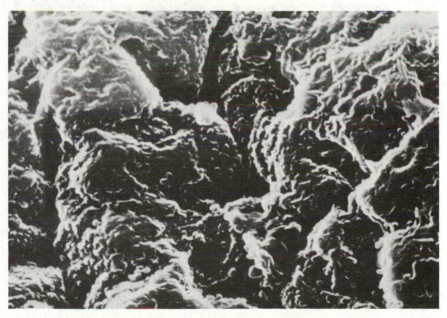

Fig. 6.7 Surface of *Artemia* larva. Note absence of micro-organisms. Bar equals 1 μm.

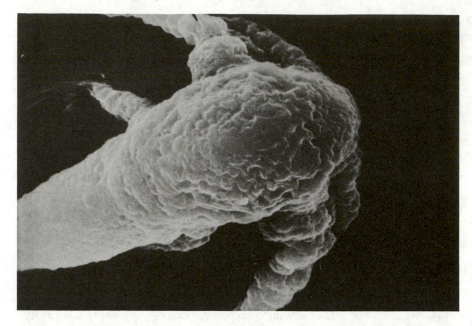

Fig. 6.8 Electron micrograph of the surface of *Artemia* larva. Bar equals 10 μm.

per nauplius. Similarly there was a lack of micro-organisms in the digestive tract of other laboratory-maintained invertebrates, e.g. the boring isopod, *Limnoria tripunctata* (Boyle and Mitchell, 1978; Sleeter *et al.*, 1978) but not so in wild-caught specimens (Unkles, 1977; Sochard *et al.*, 1979; Zachary and Colwell, 1979). As might be expected of sessile macro-organisms living in close proximity to the sediments, there is evidence from scanning electron microscopy that the shells of adult oysters (*Crassostrea gigas*) are densely covered by micro-organisms. Similarly, the exoskeleton of *Limnoria tripunctata* is extensively colonised by micro-organisms (Boyle and Mitchell, 1978), except on the mandibular apparatus and other areas of the head (Sleeter *et al.*, 1978). These micro-organisms, which mostly resembled rod-shaped and stalked bacteria, were principally located on the pleopods and telsons, with population densities estimated to be approximately one bacterium per 20 μm^2 and one bacterium per 2 μm^2, respectively. There was a dearth of eukaryotic micro-organisms, notably fungi. Moderate bacterial populations have been associated with the surface of copepods (Table 6.2) and the gills of the oyster *Crassostrea virginica* (Lovelace *et al.*, 1968). With the latter, bacterial populations ranged from 1.25×10^4 to 4.0×10^5/g. Interestingly, there was no seasonal fluctuation in numbers. Internally, the numbers of micro-organisms reflect the nature of the techniques used. Garland *et al.*

(1982) found scant evidence for the presence of attached micro-organisms on the surfaces of internal structures in adult oysters (*Crassostrea gigas*), as determined by scanning electron microscopy. Yet using plate count techniques, 3.5×10^4–2.6×10^5 bacteria/ml were present in the mantle fluid of *Crassostrea virginica* (Lovelace *et al.*, 1968). Using marine 2216E agar, Unkles (1977) recovered 4.0×10^5–2.1×10^7 bacteria/cm of gut of the sea urchin (*Echinus esculentus*). Unkles (1977) recorded 1.0×10^4–6.0×10^5 bacteria/peristomial membrane and commented on the comparative sterility of the coelomic fluid, i.e. approximately 1.6×10^3 bacteria/ml. Micro-organisms may be widely distributed throughout the digestive tract or restricted to particular niches, such as food particles. Vacelet (1975) estimated that bacteria occupied up to 40% of the volume of the mesohyl of sponges. In the case of the arctic amphipod, *Boeckosimus affinis*, Atlas *et al.* (1982) considered that the micro-organisms were in close proximity to food particles rather than attached to the gut lining. However, in the case of wild-caught *Limnoria*, Zachary and Colwell (1979) observed that bacteria are indeed associated with the gut wall, being separated by a peritrophic membrane from other microbes ingested during feeding. Numbers of gut-associated bacteria from copepods have been included in Table 6.2. Crude homogenates have been used to assess quantitative aspects of the microflora of shrimps and sponges. Thus, Ridley and Slabyj (1978) recorded 1.11×10^5 bacteria/g of freshly caught shrimp (*Pandalus borealis*). This compares to populations of 1.5×10^3–1.3×10^4 bacteria/g from pond-reared shrimps (Christopher *et al.*, 1978). From sponges, Wilkinson (1978a, b) calculated the bacterial numbers as 2.0×10^5–8.13×10^6/g.

The taxonomic composition of the microflora appears to be quite varied, largely reflecting the types present in the surrounding water. For example, Colwell and Liston (1960) determined that the natural microflora of *Cras-*

Table 6.2 *Number of aerobic heterotrophic bacteria associated with wild-caught copepods*

Animal	Surface washings[a]	Gut-associated bacteria[b]
Acartia spp.	6×10^1 –1×10^3	1.4×10^2–1×10^3
Centropages furcatus	$>10^2$	1.0×10^5
Labidocera aestiva	2×10^1–3×10^4	2.3×10^1–1×10^5
Pleuromamma spp.	1×10^3	3.0×10^2–4.7×10^2

After Sochard *et al.* (1979).
[a]Data derived from the second washing of three animals in 1.0 ml aliquots of sterile seawater.
[b]Data derived from homogenates of three animals after three washings in sterile sea water.

sostrea gigas comprised mostly *Achromobacter, Flavobacterium, Pseudomonas* and *Vibrio*. In the case of *Crassostrea virginica*, a greater diversity of bacterial taxa were recovered, including *Achromobacter, Bacillus,* coryneforms, *Cytophaga–Flavobacterium*, Enterobacteriaceae representatives, *Micrococcus, Pseudomonas* and *Vibrio* (Lovelace *et al.*, 1968). Here the bacterial taxa more closely reflected the range present in sediment, which is perhaps not surprising in view of the close proximity of the normal habitat of oysters to the benthos. Austin *et al.* (1987) identified the bacteria from bivalve larvae as *Caulobacter, Flavobacterium, Hyphomicrobium, Prosthecomicrobium* and *Vibrio*. As a result of numerical taxonomy, Wilkinson *et al.* (1981) described a range of difficult-to-identify bacteria from sponges. Here, the list included *Acinetobacter, Cytophaga–Flexibacter*, Enterobacteriaceae representatives, *Micrococcus* and *Pseudomonas. Aeromonas, Acinetobacter, Flavobacterium, Pseudomonas* and *Vibrio* were present in the sea urchin (*Echinus esculentus*) (Unkles, 1977). Other workers have highlighted the dominance of certain taxa, for example *Vibrio*, in the digestive tract. Thus, Sochard *et al.* (1979) reported the predominance of *Vibrio* and smaller populations of *Pseudomonas* in the gut of the copepod, *Acartia tonsa*. It is noteworthy that eukaryotes, particularly fungi, were rarely encountered from the samples.

Some workers have considered that the size of microbial populations within the digestive tract of invertebrates is quite low, reflecting activity by the host. This may include ciliary movement and mucus secretion, which restrict microbial growth (Garland *et al.*, 1982). Alternatively, the presence of anti-microbial mechanisms has been postulated (Unkles, 1977; Wardlaw and Unkles, 1978), notably lysozyme (Jollès and Jollès, 1975; McHenery *et al.*, 1979). Thus, it may be inferred that micro-organisms are degraded, and therefore function in the nutrition of the animals (Zobell and Landon, 1937; Zobell and Feltham, 1938; Rieper, 1978; McHenery *et al.*, 1979; Birkbeck and McHenery, 1982). At one extreme, micro-organisms may be important in the total energy input of invertebrates (Scheider and Koch, 1978), being actively ingested, or by providing certain essential nutrients, such as B-complex vitamins from bacteria or fatty acids from algae (Phillips, 1984). Rieper (1978) considered the role of bacteria as a carbon source for harpacticoid copepods. The quantity of bacteria consumed was equivalent to supplying 2.06–7.07 μg C/copepod/day. In addition, microbial enzymes may be involved in the digestion of food particles within the gastro-intestinal tract. Certainly scanning electron micrographs have clearly demonstrated the presence of microbes, notably bacteria, on food particles rather than intimately associated with the gut lining (Atlas *et al.*, 1982). However conclusive evidence for this supposition is sadly lacking.

6.4   Diseases of vertebrates

A wide range of bacteria (Table 6.3), fungi (Table 6.4), protozoa (Table 6.5) and viruses (Table 6.6) have at various times been associated with diseases or parasitic conditions in marine vertebrates. Most attention has been devoted to finfish, in farmed conditions. However it is pertinent

Table 6.3 *Bacterial pathogens of marine finfish*

Pathogen	Disease	Host range	Geographical distribution
Aeromonas salmonicida	Furunculosis	Atlantic salmon (*Salmo salar*)[a], rainbow trout (*Salmo gairdneri*), sablefish (*Anoplopoma fimbria*)	Northern Europe North America
'*Catenabacterium*[b]' spp.	—	Grey mullet (*Mugil auratus*), redfish (*Sciaenops acellata*)	USA
Eubacterium tarantellus	Eubacterial meningitis	Striped mullet (*Mugil cephalus*)	USA
Flavobacterium balustinum	Flavobacteriosis	—	USA
Flavobacterium piscicida	Flavobacteriosis	Extensive	USA
'*Flexibacter columnaris*'	Black patch necrosis	Dover sole (*Solea solea*)	UK
'*Flexibacter marinus*'	—	Black seabream (*Canthanus canthanus*), red seabream (*Chrysophorus major*)	Japan
Mycobacterium spp.	Mycobacteriosis	Extensive	Worldwide
Nocardia asteroides	Nocardiosis	Extensive	Worldwide
'*N. kampachi*'	Nocardiosis	Ayu (*Plecoglossus altivelis*), yellowtail (*Seriola quinquiradiata*)	Japan
'*Pasteurella piscicida*'	Pasteurellosis; pseudotuberculosis	Striped bass (*Morone saxatilis*), yellowtail	Japan, USA
'*Pseudomonas anguilliseptica*'	Sekiten-byo (red spot)	Eels (*Anguilla* spp.)	Japan, Scotland
Renibacterium salmoninarum	Bacterial kidney disease	Salmonids	Europe, Japan, North America
Sporocytophaga sp.	Saltwater, columnaris	Salmonids	Scotland, USA
Staphylococcus epidermidis	—	Red seabream, yellowtail	Japan
Streptococcus spp.	Streptococcicosis	Many fish species	Japan, South Africa, USA
Vibrio alginolyticus	Septicaemia	Gilthead sea bream	Israel
V. anguillarum	Vibriosis	Extensive	Worldwide
V. carchariae	Vasculitis	Sharks (*Carcharhinus* spp. spp., *Negaprion* spp.)	USA
V. damsela	Ulcer disease	Damsel fish (*Chromis puntipinnis*)	USA
V. ordalii	Vibriosis	Extensive	Worldwide
V. cholerae (non-01)	Vibriosis	Ayu	Japan
V. salmonicida	Hitra disease, (Coldwater vibriosis)	Salmonids	Scandinavia, Scotland
V. vulnificus	Vibriosis	Eels	Japan, North America

After Austin and Austin (1986).

[a] Latin names are provided for the first occurrence of a species in the list.

[b] Names in quotation marks are not included in the *Approved Lists of Bacterial Names* (Skerman *et al.*, 1980) or their supplements.

Table 6.4 *Fungal pathogens–parasites of marine finfish*

Fungal taxon	Host
Candida sp.	Trichiurids
Cladosporium sp.	Cod (*Gadus* spp.)
Cycloptericola marina	Lumpsucker (*Cyclopterus lumpus*)
Dermocystidium sp.	Salmonids
Exophiala salmonis	Salmonids
Ichthyophonus hoferi	Many fish species
Isoachyla parasitica	Sand smelt (*Atherina riqueti*)
Mycelites ossifragus	Striped wolffish (*Anarhichas lupus*)
Phialophora sp.	Atlantic salmon (*Salmo salar*)
Pullularia sp.	Sting ray (*Dasyatis* spp.)
Saprolegnia sp.	Salmonids

Modified after Möller and Anders (1986).

Table 6.5 *Protozoan parasites of marine fish*

Parasite category	Genus
Amoeba	*Entamoeba*
Ciliates	*Acanthoclinus*
	Cryptocaryon
	Gasterosteus
	Scyphidia
	Trichodina
Coccidia	*Calyptospora*
	Crystallospora
	Eimeria
	Epieimeria
Flagellates	*Amyloodinium*
	Hexamita
	Ichthyodinium
	Trypanosoma
Gregarines	*Haemogregarina*
Microsporidia	*Glugea*
	Ichthyosporidium
	Loma
	Microsporidium
	Mrazekia
	Nosema
	Plistophora
	Spraguea
	Tetramicra
Myxosporidia	*Alatosporum*
	Ceratomyxa
	Chloromyxum
	Coccomyxa
	Henneguya
	Hexacapsula
	Kudoa

Table 6.5 *(cont.)*

Parasite category	Genus
	Leptotheca
	Myxidium
	Myxobilatus
	Myxobolus
	Myxoproteus
	Myxosoma
	Parvicapsula
	Sinuolinea
	Sphaeromyxa
	Sphaerospora
	Unicapsula
	Zschokella
Sporozoan	*Ampicomplexa*

Modified after Möller and Anders (1986).

Table 6.6 *Viral diseases associated with marine fish*

Viral group	Name of virus	Host
Adenoviridae	Atlantic cod adenovirus	Atlantic cod (*Gadus morhua*)
Birnaviridae	Infectious pancreatic necrosis virus	Many species, e.g. flounder
Caliciviridae	Opaleye calicivirus	Opaleye (*Girella nigricans*)
Herpesviridae	European smelt herpes-virus	European smelt (*Osmerus eperlanus*)
	Pacific cod herpesvirus	Pacific cod (*Gadus macrocephalus*)
	Smooth dogfish herpesvirus	Smooth dogfish (*Mustelus canis*)
	Turbot herpesvirus	Turbot (*Scophthalmus maximus*),
Iridoviridae	Atlantic cod iridovirus	Atlantic cod (*Gadus morhua*)
	Lymphocystis virus	Many species
	Viral erythrocytic necrosis	Many species
Paramyxoviridae	Chinook salmon paramyxovirus	Chinook salmon (*Oncorhynchus tshawytscha*)
Reoviridae	Chum salmon reovirus	Chum salmon (*Oncorhynchus keta*)
Retroviridae	Atlantic salmon retrovirus	Atlantic salmon (*Salmo salar*)
	Northern pike retrovirus	Northern pike (*Esox lucius*),
Rhabdoviridae	Atlantic cod rhabdovirus	Atlantic cod (*Gadus morhua*)
Virus-like particles	Japanese goby papilloma virus	Japanese goby (*Acanthogobius flavimanus*)
	Flathead sole papilloma virus	Flathead sole (*Hippoglossoides elassodon*)
	Pleuronectid papilloma virus	Pleuronectidae
	Atlantic salmon hyperplasia virus	Atlantic salmon (*Salmo salar*)
	Gilthead sea bream papilloma virus	Gilthead sea beam (*Sparus auratus*)

Modified after Möller and Anders (1986)

to enquire about the precise meaning of the term disease. Although there is no acceptable definition, an appropriate attempt is:

a disease is the sum of the abnormal phenomena displayed by a group of living organisms in association with a specified common characteristic or set of characteristics by which they differ from the norm of their species in such a way as to place them at a biological disadvantage.   (Campbell *et al.*, 1979)

An interpretation of this definition is that disease is any biotic or abiotic phenomenon which causes a living organism to be abnormal or disadvantaged. According to Kinne (1980), a disease may result from the effect of pathogens, pollution, physical injury, nutritional imbalance and/or genetic disorders. Four principal types of disease are recognised in terms of epidemiology/epizootiology. These include *sporadic* outbreaks, which occur sporadically in comparatively few members of any population; *epizootics*, which are large-scale outbreaks of communicable animal disease occurring temporarily within limited geographical areas; *panzootics* which occur over wide areas; and *enzootics* which persist or re-occur as low-level outbreaks in certain areas (Kinne, 1980).

6.4.1   *Bacterial pathogens*

Representatives of the genus *Vibrio* have emerged as the scourge of marine fish. To date, eight species, including *V. alginolyticus*, *V. anguillarum*, *V. carchariae*, *V. cholerae*, *V. damsela*, *V. ordalii*, *V. salmonicida* and *V. vulnificus* have been described as fish pathogens. Generally, the organisms cause septicaemic conditions involving the skin and internal organs for which the name of vibriosis has been coined. Lethargy may be apparent. The organisms may be readily isolated from infected tissue by means of seawater agar (e.g. marine 2216E agar) and/or thiosulphate citrate bile salt sucrose agar (TCBS) with incubation at 15–25 °C for approximately 48 hours. The taxonomy of the organisms is comparatively well understood (Austin and Austin, 1986), and fresh isolates are readily identified by means of phenotypic tests and diagnostic tables. The pathogenic mechanisms involve exotoxin (exo-enzyme) production, notably proteases and haemolysins. Generally, the organisms occur in the marine and estuarine environments, from which they initiate infection cycles.

Vibrio anguillarum has attracted considerable attention, largely because of the havoc caused in fish farming operations. The excellent description of an outbreak of 'red-pest' in eels during 1718 is undoubtedly the first reference to a bacterial fish disease in the European literature (Bonaveri, 1761). The taxonomy of the pathogen had a chequered history, which led to the division of the taxon into two separate species, i.e. *V. anguillarum* and *V. ordalii* (Schiewe, 1981). Historically, the problems began with the recognition of two biotypes of *V. anguillarum* by Nybelin (1935). These were labelled as Type A (*Vibrio anguillarum forma typica*) and Type B (*V. anguillarum forma anguillicida*). Type C (*V. anguillarum forma opthalmica*) was described by Smith (1961) to accommodate Japanese strains which had been labelled as *V. piscium* var. *japonicus* (David, 1927; Hoshina, 1956). However the relationship of these biotypes with *V. ichthyodermis* (Wells and Zobell, 1934) and *V. anguillicida* (Nishibuchi and Muroga, 1977) awaits clarification. Serology has further complicated the understanding of *V. anguillarum* insofar as the establishment of serotypes has transversed phenetic boundaries. Initially three serotypes were established (Pacha and Kiehn, 1969), although this number was subsequently increased to six (Kitao *et al.*, 1983).

Although *V. anguillarum* is part of the normal microflora of water (Larsen, 1982) and fish (Mattheis, 1964), the precise mode of infection is unclear, but probably involves attachment to and then colonisation of the host, followed by penetration of the tissues. It has been postulated that infection may begin with colonisation of the posterior gastro-intestinal tract and rectum (Ransom, 1978). Resistance to the potentially debilitating effect of fish serum may hasten the invasion process (Trust *et al.*, 1981). A plasmid, designated pJM1 (of 47×10^6 daltons molecular weight), has been implicated with virulence (Crosa *et al.*, 1977, 1980; Crosa, 1980). The role of this plasmid concerns specifying an iron-sequestering mechanism, which enables the bacterial cell to compete for available iron in the fish tissues. This system encompasses a low molecular weight siderophore (a biological chelating agent) and two outer membrane proteins, designated as OM2 (molecular weight = 86 000 daltons) and OM3 (molecular weight = 79 000 daltons). It has been established that the siderophore and outer membrane protein, OM2, are coded by plasmid pJM1, whereas OM3 is a function of chromosomal involvement. The mechanism involves diffusion of the siderophore into the environment, and the formation of iron complexes which attach to OM2 and lead to transport of the iron into the bacterial cell (Crosa and Hodges, 1981; Crosa *et al.*, 1983; Tolmasky and Crosa, 1984; Actis *et al.*, 1985). Thus invading bacteria may multiply in the host by scavenging successfully for the iron which is bound by high-affinity iron-binding proteins such as transferrin, lactoferrin and ferritin. These are present in the serum, secretions and tissues, respectively (Bullen *et al.*, 1978). Certainly *V. anguillarum* has been one of the few successful candidates for vaccine development. Commercial formalin-inactivated preparations are available which have gained widespread use in aquaculture. The immunogenicity of the pathogen reflects the presence of heat-stable lipopolysaccharides in the cell wall, which may be released in the culture supernatant (Chart and Trust, 1984). These large molecular weight compounds (100 000 daltons) confer protection on the recipient host (Evelyn and Ketcheson, 1980).

A second scourge of marine fish is represented by the genus *Mycobacterium*. The first report of acid-fast bacteria in marine fish was published by von Betegh (1910). This was followed by the isolation of *Mycobacterium marinum* from the liver, spleen and kidney of tropical coral fish in 1926 (Aronson, 1926). The first case of 'tuberculosis' in wild fish was reported from cod (*Gadus morhua*), which was landed at Fleetwood (Alexander, 1913). To date mycobacteriosis has been observed in >150 species of marine and freshwater fish (Nigrelli and Vogel, 1963), and the aetiological agents have been classified in a wide assortment of species, including *M. anabanti, M. chelonei* subsp. *piscarium, M. fortuitum, M. marinum, M. piscium, M. platypoecilus, M. ranae* and *M. salmoniphilum*. Mycobacteriosis is a chronic progressive disease with various external signs including emaciation, inflammation of the skin, exopthalmia (bulging eyes), open lesions and ulceration. Internally, grey-white nodules (granulomas) develop on organs, notably the liver, kidney, heart and spleen (Van Duijn, 1981). The disease may take several years to progress from the asymptomatic state to the clinical illness. Attempts at isolating the pathogen often fail, indicating fastidious nutritional requirements. Nevertheless, some success has been achieved with Petragnani, Löwenstein-Jensen, Middlebrook 7H10 and Dorset Egg media with incubation at 15–22 °C for 2–28 days (Dulin, 1979). Unfortunately, most descriptions of fish pathogenic mycobacteria are poor, and it is often difficult to determine whether isolates belong in *Mycobacterium* or *Nocardia*. Very little is known about the ecology of the organisms. However it seems likely that the reservoir of infection is the aquatic environment, although it is unknown what factors lead to the development and spread of the disease.

6.4.2   Fungi

The limited attention focused on 'marine' fungi is reflected by the inadequate knowledge of fungal diseases of marine vertebrates. The list of aetiological agents is small (Table 6.4), including organisms of notoriety in freshwater conditions, e.g. *Dermocystidium, Isoachyla* and *Saprolegnia*. These are likely to be problematic only in coastal sites which receive freshwater run-off particularly from rivers. Such organisms may colonise compromised hosts, covering them with a visible mycelium. Limited reports have been made of the pathogenic role of presumptive *Cladosporium, Cycloptericola* and *Pullularia* in North Sea cod (Reichenbach-Klinke, 1954), lumpfish (*Cyclopterus lumpus*) (Apstein, 1910) and aquarium sting ray (*Dasyatis* spp.) (Otte, 1964), respectively. Consequently the relevance of these fungal genera to fish pathology is uncertain. The dermatiaceous hyphomycetes, *Exophiala* and *Phialophora* have been occasionally trouble-some in farmed populations of channel catfish (*Ictalurus punctatus*) (McGin-nis and Ajello, 1974) and Atlantic salmon (*Salmo salar*) (Ellis *et al.*, 1983). In particular 10% mortality among Atlantic salmon in a Scottish fish farm was attributed to *Exophiala* (Richards *et al.*, 1978). More serious losses, of up to 50%, occurred in Norway (Langvad *et al.*, 1985). A poorly defined group of fungi, referred to as *Mycelites ossifragus*, affect the teeth of rays (*Dasyatis* spp.) and wolffish (*Anarhichas* spp.), and cause a condition akin to dental caries (Peyer, 1926; Kerebel *et al.*, 1979). Presumptive *Candida* have been implicated with systemic infections in trichiurids (Grabda, 1982).

The best documented fungal pathogen/parasite is *Ichthyophonus hoferi* (see Neish and Hughes, 1980; McVicar, 1982), which has been found in approximately 80 species of fish worldwide (Reichenbach-Klinke, 1954, 1955). In some geographical areas high incidences of ichthyophonus disease have been reported. For example, 15% of cod were estimated to be affected in the west Baltic Sea during 1972 (Möller, 1974). Indeed the disease may devastate wild fish populations, with death ensuing within 30 days of the initial infection. External signs of disease include raised scales, emaciation, and impaired swimming, i.e. tumbling. Internally, granulomas may occur in the muscle. The principal organs affected include the heart, kidney, liver and spleen. Infection commences by the ingestion of the heavy-walled spores, which germinate in the stomach with the release of amoeboid cells. These penetrate through the wall of the gastro-intestinal tract, and are dispersed through the blood stream. Thereafter encapsulation of the fungus may occur in the body tissues. Alternatively, multi-nucleated plasmodia may be produced, which divide and continue the invasion processes.

6.4.3 *Protozoa*

A substantial number of protozoan genera, including amoebae, coccidia, flagellates, gregarines, microsporidia, myxosporidia and sporo-zoans, have been found to parasitise marine fish (Lom, 1984) (Table 6.5). Indeed, most marine fish harbour at least one species of protozoan parasite, although the relevance to disease is often obscure. Infestation by *Cre-pidoodinium* may cause gill hyperplasia, although the condition is not fatal. However mass mortalities may result from infection with trypanosomes. For further information on protozoan parasites the reader should consult Möller and Anders (1986).

6.4.4   *Viruses*

Approximately 20 viruses, representing the Adenoviridae, Birnaviridae, Caliciviridae, Herpesviridae, Iridoviridae, Paramyxoviridae, Reoviridae, Retroviridae and Rhabdoviridae, have been seen and/or isolated from marine fish (Table 6.6). However, the significance of some of these viruses to disease pathology is unclear. In most other cases they are associated with neoplastic–hyperplastic conditions. An adenovirus (icosohedral; 77 nm in diam.) from Atlantic cod (*Gadus morhua*) has been observed by electron microscopy from cases of epidermal hyperplasia (Jensen and Bloch, 1980). Yet isolation of the virus was not reported. An iridovirus (145–150 nm in diam.) and a rhabdovirus (55 × 175 nm in size) have been isolated from skin diseases in Atlantic cod (Jensen *et al.*, 1979). Nevertheless, the significance of these viruses to disease pathology is unclear. In addition, viral-like particles have been observed in papillomas (Table 6.6), although the taxonomic status and role of the objects remain to be resolved.

Five virus diseases have received great attention:

Chum salmon reovirus. This is an icosohedral, double stranded RNA virus, which was recovered from adult chum salmon (*Oncorhynchus keta*) returning to a hatchery in Hokkaido, Japan (Winton *et al.*, 1983). The virus may be replicated in chinook salmon (*Oncorhynchus tshawytscha*) cell lines at 10–20 °C. The genome, in 11 segments, has been determined to have a molecular weight of 16×10^6 daltons. There are five major proteins, of between 34 000 and 137 000 daltons molecular weight. The viral particles are 75 nm in diameter, with double capsids. Resistance has been demonstrated to pH 3, although inactivation occurs by heating to 56 °C. Taxonomically, the virus was considered to resemble the orthoreovirus group. In experimentally infected fish, <5% mortalities were recorded.

Herpes virus. There have been several reports of herpes virus infections of marine fish species, including Pacific cod (*Gadus macrocephalus*) (McCain *et al.*, 1979) and turbot (*Scophthalmus maximus*) (Buchanan and Richards, 1982). The virus appears to affect the epidermis and gill epithelia, causing the formation of giant cells and intranuclear inclusions. In the case of *Herpesvirus scophthalmi*, hexagonal particles of 100 nm in diameter have been observed by electron microscopy, although replication of the virus in tissue culture has not been achieved (Buchanan and Madeley, 1978; Buchanan *et al.*, 1978; Buchanan and Richards, 1982). Initially infected fish became lethargic and ceased feeding. Heavy losses of approximately 30% occurred in juvenile turbot during one outbreak in a Scottish fish farm.

Infectious pancreatic necrosis virus. Although primarily troublesome in juvenile freshwater fish, this birnavirus has been associated with disease outbreaks in marine fish including flounder (*Pseudopleuronectes* spp.) (McAllister *et al.*, 1983), sea bass (*Dicentrarchus salmoides*) (Hill, 1982) and eels (*Anguilla* spp.) (Sano *et al.*, 1981). Generally, the virus (icosohedral, 55–60 nm in diam.) replicates well in tissue culture.

Lymphocystis virus. First described in flounder (*Pseudopleuronectes* spp.) by Lowe (1874), lymphocystis has emerged as a widespread and serious condition of marine fish. The condition is extremely infectious and has, to date, been recognised in 49 species of fish, where up to 33% of individuals in some populations may be affected. The virus causes the formation of white nodules on the body and fins. It would appear that it varies in size from 130–300 nm. Smaller sized virions of 130–150 nm have been observed in pleuronectids, whereas particles of 300 nm have been recorded in herring (Aneer and Ljungberg, 1976).

Viral erythrocytic necrosis virus. This condition was initially described by Laird and Bullock (1969) as concerning the erythrocytes in the blood of coastal marine fish. Eosinophilic inclusion bodies of 1 μm in diameter occurred in the cytoplasm of the erythrocytes, whereas the nuclei were distorted and developed vesicles containing densely stained particles of 250–500 nm in size. The aetiological agent has not been replicated in tissue culture.

6.5 Diseases of invertebrates

Most species of invertebrates will at one time or another suffer the rigours of disease. Numerous microbial pathogens and parasites have been implicated, including algae, bacteria, fungi, protozoa and viruses (Table 6.7). Detailed coverage has been provided by Fisher *et al.* (1978) and Leibovitz (1978). Some of these micro-organisms have gained notoriety, e.g. *Vibrio anguillarum*, whereas others have limited significance in pathology, e.g. the poorly described *Achromobacter* sp. Species such as *V. anguillarum* are troublesome among many groups of invertebrates from wide geographical areas. Others, e.g. *Streptococcus faecalis* subsp. *liquefaciens*, appear to be restricted to one host with limited geographical distribution. With few exceptions, the biology of the aetiological agents is incompletely understood. The exceptions require further discussion.

6.5.1 'Vibriosis'

Tubiash *et al.* (1965) may be credited with the pioneering work on vibriosis in oysters. The name of the disease has become established in the published literature, but is now associated with a diversity of causal agents in a wide range of invertebrate species. The common link is that the pathogens are bacterial, and inevitably belong to the genus *Vibrio*. The principal

Table 6.7 *Microbial pathogens and parasites of marine invertebrates*

Pathogen/parasite	Host and disease name
Algae	
Anabaena sp.	Shrimps (*Penaeus* spp.)[a]
Oscillatoria sp.	Shrimps
Bacteria	
Achromobacter sp.	Oysters (*Ostrea* spp., *Crassostrea* spp.)
Aerococcus viridans subsp. *homari*	Lobsters (gaffkemia) (*Homarus* spp.)
Aeromonas sp.	Oysters
Alcaligenes faecalis subsp. *homari*	Lobsters
Alteromonas sp.	Oysters, shrimps (shell-disease)
Chlamydia spp.	Clams (*Mercenaria mercenaria*)
Cladothrix dichotoma	Oysters
Coxiella sp.	Tellina (*Tellina tenuis*)
Leucothrix mucor	Lobsters, shrimps (shell-disease)
Nocardia sp.	Oysters
Photobacterium sp.	Tanner crabs (*Chionoecetes tanneri*)
Pseudomonas enalia	Oysters (gaping disease)
Spirillum sp.	Shrimps (shell-disease)
Spirochaete	Artemia (*Artemia salina*)
Streptococcus faecalis subsp. *liquefaciens*	Crabs (*Carcinus* spp., *Cancer* spp.)
Vibrio alginolyticus	Albalone (*Haliotis* spp.) clams, oysters, (vibriosis), shrimps
V. anguillarum	Abalone, clams, oysters (vibriosis)
V. parahaemolyticus	Crabs, shrimps
V. pelagia	Oysters, red abalone (*Haliotis* sp.)
V. splendida	Oysters
Vibrio sp.	Clams, lobsters, (shell-disease), oysters (vibriosis), shrimp (shell-disease)
Fungi	
Dermocystidium marinum	Oysters
Didymaria palinuri	Lobsters
Fusarium solani	Lobsters, prawns (*Penaeus* spp.) (black gill disease)
Haliphthoros milfordensis	Lobsters
Lagenidium callinectes	Crabs
Leptolegnia marina	Crabs
Plectospira dubia	Crabs
Pythium thalassium	Crabs
Ramularia branchialis	Lobsters
Sirolpidium zoophthorum	Oysters
Protozoa	
Ancistrocoma pelseneeri (ciliate)	Oysters, mussels (*Mytilus* spp.)
Ancistrocoma myae (ciliate)	Clams
Ancistruma mytili (ciliate)	Mussels
Anophrys sarcophaga (ciliate)	Crabs
Bonamia sp.	Oysters
Carcinoecetes conformis (gregarine)	Crabs
Cephaloidophora spp. (gregarine)	Crabs
Cephalolobus petiti (gregarine)	Shrimps
Chytridiopsis mytilovum (haplosporidian)	Mussels

Table 6.7 *(cont.)*

Pathogen/parasite	Host and disease name
C. ovicola (haplosporidian)	Oysters
Ephelota gemmipara (suctorian)	Lobsters
Epistylus spp.	Crabs, shrimps
Flabellula spp. (amoeba)	Oysters
Haplosporidium tumefacientis (haplosporidian)	Mussels
Hematodinium sp. (ciliate)	Crustacea
Hexamita nelsoni (flagellate)	Oysters (hexamitiasis)
Hypocomides mytili (ciliate)	Mussels
Kidderia mytili (ciliate)	Mussels
Lagenophrys sp. (ciliate)	Crabs
L. lunatus (ciliate)	Shrimps
Marteilia sp.	Oysters
Minchinia costalis (haplosporidian)	Oysters
M. louisiana (haplosporidian)	Crabs
M. nelsoni (haplosporidian)	Oysters
Nematopsis duoriar (gregarine)	Shrimps
N. ostrearum (gregarine)	Oysters
N. penaeus (gregarine)	Shrimps
N. prytherchi (gregarine)	Oysters
N. schneideri (gregarine)	Mussels
Nosema nelsoni (microsporidian)	Shrimps
Nosema pulvis (microsporidian)	Crabs
Orchitophrya stellarum (ciliate)	Starfish (*Asterias* spp.)
Paradinium sp. (ciliate)	Copepods
Paramoeba sp.	Crustacea
Plistophora cargoi (microsporidian)	Crabs
Porospora gigantea (gregarine)	Lobsters
P. nephropis (gregarine)	Lobsters
Sphenophrya sp. (ciliate)	Oysters
Syndinium sp.	Copepods
Teresbrospira lenticularis (ciliate)	Shrimps
Thelohania duorara (microsporidian)	Shrimps
T. maenadis (microsporidian)	Crabs
Trichodina myicola (ciliate)	Clams
Vorticella sp.	Shrimps
Viruses	
Birnavirus	Tellina
Herpesvirus	Crabs, oysters
Iridovirus	Cephalopods, oysters
Ovacystis virus	Oysters
Papovavirus	Oysters
Reovirus	Cuttlefish (*Sepia* spp.), oysters
Retrovirus	Clams (infectious hematopoietic neoplasm)

[a]Latin names are provided for the first occurrence of a species in the list.

pathogen is *V. anguillarum* (e.g. Brown, 1981) (the role of the organism formerly known as *V. anguillarum* biotype II = *V. ordalii* is unclear) although there is widespread trouble with *V. alginolyticus* (Elston and Lockwood, 1983) and *V. parahaemolyticus* (Vanderzant and Nickelson, 1970) and limited involvement of *V. pelagia* (Elston and Lockwood, 1983) and *V. splendida* (Jeffries, 1982). Unfortunately an opinion may be reached that any case of Gram-negative bacterial disease will be labelled as vibriosis, with *V. anguillarum* as the supposed causal agent. This need not be strictly true, but reflects sampling and identification techniques. It is unresolved whether *V. anguillarum* should be considered as a primary pathogen, capable of devastating healthy stocks, or if the organism is troublesome in compromised hosts. Although published information supports the former, personal experience suggests the latter.

The first stage of the disease process is availability of the pathogen, followed by association with the host, and thence expression of the disease. Vibrios are common in the aquatic environment, and, therefore there is a constant source of inoculum around the invertebrates. Evidence points to the preferential attachment of pathogen cells to shells, particularly of larval animals, e.g. *Crassostrea virginica* (Elston and Leibovitz, 1980; Elston *et al.*, 1981). Early signs of distress in larval oysters include loss of motility and abnormalities in the velum (Elston and Leibovitz, 1980; Garland *et al.*, 1983). The velar cells become detached or disorganised and lose their retractor muscle insertions. In more advanced larvae, the earliest signs of disease involved detachment of absorptive cells of the digestive gland (Elston and Leibovitz, 1980). This leads to interference with digestive processes (Elston *et al.*, 1982), and hence loss of nutrient-utilising capability. Elston *et al.* (1981) reported the deposition of lipids and the formation of large vacuoles in the digestive gland of oysters (*Crassostrea virginica*). In oysters (*Crassostrea virginica*, *Ostrea edulis*) and clams (*Mercenaria mercenaria*) the bacteria also interfered with the deposition of calcium and protein in the shell (Elston *et al.*, 1982). This led to shell abnormalities. The pathogenicity mechanism appears to involve water-soluble heat-stable exotoxins, which inhibit larval swimming (Di Salvo *et al.*, 1978).

Sudden mass mortality among cultured bivalves, attributed to vibriosis and reflective of increasing water temperatures (Grischkowsky and Liston, 1974), is a recurring problem in hatcheries worldwide (e.g. Garland *et al.*, 1983). It is contended that animals which have been stressed or weakened by unfavourable water quality conditions would be more susceptible to disease. Under these conditions, the bacteria should be regarded as opportunistic invaders rather than as primary pathogens. Helm (1971) implicated algae, notably *Phaeocystis*, with causing deleterious effects in oyster larvae.

Cardwell (1978) related toxicity in larvae to metabolites from plankton. Moreover, Elston *et al.* (1981) suggested that poor water quality, in particular the presence of pesticides and petroleum products, may induce vibriosis. The involvement of food in the transmission of pathogens has been well documented (Elston *et al.*, 1982; Garland *et al.*, 1983; Austin *et al.*, 1987). Thus high counts of oyster pathogens have been recovered from algal food cultures, namely *Chaetoceros* and *Isochrysis*. According to Garland *et al.* (1983), the bacterial counts on the algal food cultures may exceed 10^7/ml (*Chaetoceros* = 8×10^7 bacteria/ml, *Isochrysis* = 3×10^7 bacteria/ml). Addition of such large bacterial populations to weakened animals would inevitably lead to dire consequences. In the case of *Crassostrea gigas*, Garland *et al.* (1983) reported a hundred-fold increase in the bacterial numbers from healthy to inactive larvae. With four-day-old larvae, there was an increase from 10^4 bacteria/1000 larvae in normal situations to 1.5×10^6 bacteria/1000 larvae in inactive groups. Similarly with 9-day-old animals, the bacterial counts rose from 2.5×10^5/1000 larvae (normal) to 5×10^7/1000 larvae (inactive).

Seemingly, the full story about vibriosis has still to be unravelled. Further detailed work is necessary.

6.5.2   Shell-disease

Shell-disease of lobsters was initially reported by Hess (1937), and was believed to be caused by chitinolytic bacteria. Thereafter, Rosen (1970) implicated fungi. The condition is now recognised among many groups of invertebrates, including shrimps and crabs. Essentially, there is a progressive softening and pitting of the exoskeleton, which is usually accompanied by darkening of the necrotic area. An example is the so-called brown spot disease of Gulf coast shrimp (*Penaeus* sp.). Generally, dense populations of bacteria, which degrade chitin, lipid and protein, are present in the diseased exoskeleton. As far as may be ascertained the organisms do not spread into underlying tissues. Instead the necrotic areas may serve as portals of entry for secondary invaders. The causal agents include *Alteromonas, Photobacterium, Spirillum* and *Vibrio*. Indeed, from the deep sea tanner crab (*Chionoecetes tanneri*), Baross *et al.* (1978) recovered *Photobacterium*, which degraded chitin at *in situ* temperatures. By virtue of its widespread presence this organism was presumed to be involved with the disease.

6.5.3   Gaffkemia

This is an economically serious disease of lobsters, which appears to have been initially recognised in the years immediately after the end of the Second World War (Hitchner and Snieszko, 1947; Snieszko and Taylor, 1947). The causal agent had a chequered taxonomic history, being originally classified as *Gaffkya homari*, then re-classified as *Pediococcus homari* before transfer to *Aerococcus*, as *Aerococcus viridans* subsp. *homari*. Cultures contain fairly unreactive Gram-positive large cocci, forming distinctive tetrads (Fig. 6.9). The catalase-negative, α-haemolytic cells may be cultured on phenyl ethyl alcohol blood agar (Stewart *et al.*, 1966; Kellogg *et al.*, 1974) with incubation at 28 °C for 24–48 hours, whereupon small (approximately 1 mm diam.), flat, grey colonies develop. Selective enrichment has also proved feasible in cell cultures (Table 6.8; Stewart, 1972). The organism is regarded as non-invasive, entering the host through wounds whereupon rapid multiplication ensues, such that up to 10^8 bacteria/ml of haemolymph may result. This is matched by a concomitant decrease in the number of circulating haemocytes and quantity of sugars in the haemolymph (Stewart and Cornick, 1972). External signs of gaffkemia include lethargy and cessation of feeding. Treatment of early cases of disease is possible by injection of vancomycin (Stewart and Arie, 1974). Prevention of outbreaks may result from improvements in water quality. Available evidence suggests that the organism comprises part of the normal microflora on the exoskeleton of lobsters. In addition, the aerococci may be readily recovered from marine

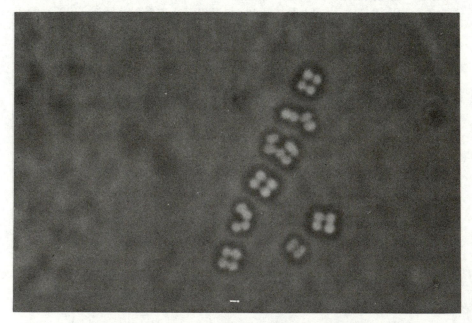

Fig. 6.9 Phase-contrast micrograph of tetrads of *Aerococcus viridans* subsp. *homari* in the haemolymph of lobster. Bar equals 1μm.

sediments in the vicinity of lobsters (Kellogg *et al.*, 1974). Although natural infections are mostly restricted to European and American lobsters, other crustacea, e.g. crabs, succumb to artificial challenge (Rabin and Hughes, 1968).

6.5.4 *Epibiotic associations*

Microbial epibionts, particularly on gills and eggs, may be debilitatory to invertebrates, resulting in asphyxia (Johnson, 1974) or interference with moulting processess. The most commonly encountered epibiont is the filamentous bacterial taxon, *Leucothrix mucor* (Johnson *et al.*, 1971). This organism may be present as strands of 2 μm in diameter by 14–15 μm in length. Algae of the genera *Anabaena* and *Oscillatoria* may appear as 10–15 μm wide × >1 mm long strands, particularly on eggs. These algae are especially harmful to larvae during moulting. Algae of the family Rivulariaceae and stalked protozoans, namely *Epistylis* and *Vorticella*, have also been implicated.

6.5.5 *Fungal diseases*

Several fungal genera have gained notoriety as causal agents of serious disease conditions in invertebrates. A deuteromycete of the genus *Fusarium*, probably *Fusarium solani*, has been associated with a gangrenous-type condition in the appendages, gills and in the exoskeleton of lobsters (Lightner and Fontaine, 1973). The lesions appear as expanding cuticular or sub-cuticular zones, which darken due to the deposition of melanin.

Table 6.8 *Selective enrichment medium for* Aerococcus viridans *subsp* homari

Component	Proportion (% w/v)
Glucose	0.65
Phenyl ethyl alcohol	0.25
Sodium chloride	0.64
Tryptone	1.5
Yeast extract	0.45
Brom cresol purple	0.0008
Experimental conditions; pH 7.4	

After Stewart (1972).
Incubation is for 5 d at 25°C, whereupon a positive result is indicated by a colour change from purple to yellow.

Occasionally they contain red decomposing centres because of the activity of toxins.

The phycomycete, *Haliphthoros milfordensis*, also effects lobsters (Fisher *et al.*, 1975). The slightly chitinolytic laterally biflagellate, free-swimming zoospores penetrate through small wounds, whereupon the site of infection assumes a reddish-brown colour. Moulting may be hindered. The sodium-requiring organism may be readily cultured on cornmeal dextrose agar.

Another phycomycete, *Lagenidium callinectes*, was reported by Couch (1942) as parasitic on the eggs of the blue crab (*Callinectes sapidus*). Zoospores are formed, which swim directly towards the eggs.

6.5.6   *Virus diseases*

Overall there is little information available on viral diseases of marine invertebrates, perhaps reflecting a general absence of suitable cell-lines and thus difficulty with isolation of many of the offending pathogens. The first report of a virus disease of marine invertebrates stems from the work of Vago (1966). Now, ten viruses are recognised among crustaceans, with eight being derived from decapods and one each from crabs (*Carcinus, Cancer* spp.) and pink shrimps. Hill (1984) discussed the presence of virus-like particles in 14 species of molluscs, although sometimes there was a complete absence of pathology. The groups of viruses described to date include birnavirus, herpesvirus, iridovirus, papovavirus and reovirus. Hill (1976) grew a birnavirus from the mollusc, *Tellina tenuis*, in tissue culture. This virus, which had a diameter of only 59 nm, sedimented at 430 Svedberg units (S) in sucrose gradients, and banded at a density of 1.32 g/ml in caesium chloride gradients. RN'ase-resistant RNA, with a molecular weight of approximately 2.8×10^6 daltons, and two major polypeptides (40×10^3 daltons and 67×10^3 daltons molecular weight) and one minor polypeptide of 110×10^3 daltons molecular weight were present (Underwood *et al.*, 1977). This virus was described as a serotype of the finfish pathogenic infectious pancreatic necrosis virus (IPNV) group. To date, this virus would appear to be the best studied of all marine invertebrate viruses.

Farley *et al.* (1972) have been credited with the first description of herpes-like particles in American oysters (*Crassostrea virginica*). Mortalities were temperature-dependant, amounting to 52% of the animals at a water temperature of 28–30 °C. Internally, signs of disease were indicated by the presence of dilated digestive diverticulae. Within the cells, intra-nuclear inclusions were observed, which contained single membrane-coated hexagonal viral particles of 70–90 nm in diameter. A second disease, attributed to herpesvirus, was recorded in blue crabs (*Callinectes sapidus*). Here, infected animals contained milky haemolymph. Enveloped viral particles of 185–214 nm in diameter were observed. So far, the herpesvirus diseases appear to be related to conditions of overcrowding and temperature-mediated stress.

Interest in iridovirus infections of oysters, e.g. gill disease of the Portugese oyster (*Crassostrea angulata*), started in 1966 following an outbreak of disease in France in which mortality levels reached 40%. The condition began with the appearance of small yellow spots and ulcerations on gills and labial palps. Gradually, these lesions enlarged to form brown necrotic areas throughout the branchial lamellae. Ultimately, there was total destruction of the gill and, hence, death. Giant globular cells, of 30–40 μm in size were observed. These contained basophilic, Feulgen-positive cytoplasmic inclusions of 3–4 μm in diameter and many granules (0.3 μm in size). The basophilic inclusions comprised hexagonal viral particles of 300 nm in diameter. A second iridovirus was associated with a tumour condition in captured octopus (*Octopus vulgaris*) from Italy (Rungger *et al.*, 1971). In these specimens, hexagonal viral particles (120–140 nm $\times$ 95–105 nm) were observed.

A papova-like virus was associated with a pathology in gonad tissue of *Crassostrea virginica* (Farley, 1976). In particlar, the gonad contained abnormally large cells of 500 μm. These contained negligible cytoplasm. However, the nuclei, which contained many large (5 μm) nucleoli, appeared granular and strongly Feulgen-positive. Intranuclear inclusions, which were also Feulgen-positive, comprised hexagonal or pentagonal virus particles of 53 μm in diameter. This morphology was indicative of papillomavirus (Hill, 1984).

Some information has been published about reovirus in cuttlefish (*Sepia officinalis*) (Devauchelle and Vago, 1971) and *Crassostrea virginica* (Meyers, 1979). In the former example, viral-like particles of 75 nm in diameter were observed in stomach tissue.

Finally, a retrovirus has been associated with infectious hematopoietic neoplasm in the soft shell clam, *Mya arenaria*. With this condition, there was a proliferation of atypical cells and the ultimate destruction of normal host cells. From infected animals, Cooper *et al.* (1982) described an enveloped pleomorphic virus of 120 nm with eccentric nucleoids of 60–70 nm.

6.5.7 *Rickettsial diseases*

So far, there has been only one documented case of a rickettsial disease among marine invertebrates. Thus, Buchanan (1978) described a disease of the digestive gland of tellina (*Tellina tenuis*), from which a rickettsial-like organism was cultured in embryonated hens eggs at 37 °C. The organism, of 300–500 nm in diameter and considered to resemble *Coxiella*, was spherical to coccoid in shape, paired, possessed a double membrane, and was rich in ribosomes. Bacteriophage may have been present within the cells of the pathogen. The initial sign of infection was the presence of small eosinophilic inclusion bodies in the region normally occupied by the golgi apparatus. These inclusion bodies enlarged, and eventually burst.

7 Microbiology of the deep sea

Vivid descriptions of the deep sea were portrayed in the fictional film about raising the Titanic from her grave in the Atlantic Ocean. Yet who would have realised that only a few years later real photographs would be obtained of the vessel in her watery location. The public's curiosity must surely have been aroused by photographs of sections of the hull, and of wine bottles which appeared to be in a remarkable state of preservation after 80 years of submersion in that cold seemingly inhospitable environment. Could life really exist in these deep waters? In fact, this begs the question of how deep is deep? An appropriate answer is that deep is defined as meaning depths exceeding 1000 m. To put this into perspective, 88% of the area of the sea could, therefore, be deemed as 'deep'. Alternatively, 75% of the total volume of the oceans comprises deep sea. Studies concerning the microbiology of the deep sea have examined a variety of depths, from 1450 m (Jannasch *et al.*, 1971) to 10 480 m (Zobell and Morita, 1959).

One and a half centuries ago, it was assumed that there was no life below a depth of approximately 600 m. This was the opinion of the British oceanographer, Edward Forbes, in the 1840s. Nevertheless, three decades later the science of deep-sea biology was truly borne during the Challenger Expedition of 1873–6. Deep-sea bacteria were recovered by Certes (1884a) as a result of the Travaillier and Talisman Expeditions of 1882–3. Certes recovered balotolerant (pressure tolerant) bacteria from depths of 5000 m, and suggested that such micro-organisms may exist in a state of suspended animation (Certes, 1884b). This concept has relevance for marine microbiology in general, and has been discussed at length separately (see Chapter 5). With the discovery of agar as a gelling agent for microbiological media, its use was quickly adopted in studies of the deep sea. Thus during a trans-Atlantic crossing by a passenger ship in 1886, few colony-forming units resulted from inoculation with samples of water obtained from depths exceeding 1110 m (Fischer, 1894). However, this lack of growth may have reflected nutritional deficiences in the composition of the media. Subsequent improvements in the knowledge of deep-sea biology slowly occurred, but it was only after the end of the Second World War that progress rapidly ensued thanks largely to the efforts of Zobell and co-workers. This group initiated work on the effect of hydrostatic pressure on bacterial activity (Zobell and Johnson, 1949). Thence, participation in the Danish Galathea Expedition in 1950–2 started Zobell on his pioneering work in deep-sea microbiology. For useful reviews of deep-sea microbiology readers are directed to Morita (1980), Jannasch and Taylor (1984) and Jannasch (1985).

7.1   The deep-sea environment

The impression may be gained that the deep sea is a dark inhospitable environment which is devoid of food and life. Instead, food reaches the bottom of even the deepest trenches, where it supports an abundance of intriguing life forms. Apart from the sporadic arrival of large chunks of food, such as dead fish (these settle at rates of 50–500 m/h, and do not degrade greatly *en route* to the sea floor), there is a more continuous settling to the sea floor of minute organic particles. These comprise the so-called 'marine snow' (Silver *et al.*, 1978). This includes highly-recalcitrant molecules and faecal matter as well as components of the phytoplankton and zooplankton. It has been estimated that phytoplankton sink at a rate of only 0.1–1.0 m/day (Vinogradov and Tseitlin, 1983), but nevertheless they constitute a constant source of food in the deep sea. In comparison the sinking rates for faecal pellets and zooplankton (salps) is 50–200 m/day and up to 2700 m/day, respectively. Menzel and Ryther (1970) considered that the distribution of particulate and dissolved organic matter, distinguished according to retention or passage through a 1.2 μm pore filter (through which bacteria will pass, and therefore be considered soluble), was homogeneous below depths of approximately 200–300 m.

Certainly it is readily admitted that the deep sea is hardly a nutrient-rich environment compared to the surface waters, and it remains unclear what the dominant source of nutrients is. Yet this lower level of nutrients must favour the survival of oligotrophs; a contention which is readily supported by all the available literature. More specifically, the deep sea seems to have a particularly low input of organic carbon, with only approximately 1–5% of the photosynthetic organic carbon reaching the bottom (Suess, 1980). Honjo (1978) estimated the rate of decomposition of organic carbon in deep water to be approximately 2.2 mg $C/m^2/day$. This value coincides with rates of oxygen consumption, as determined by other workers.

Most of the deep sea is cold i.e. at 3 °C $\pm$ 1 °, (the obvious exception of warm and hot hydrothermal vents will be discussed separately) which may favour the presence of psychrophiles. The deep sea is also under great pressure (approximately 1 atmosphere of pressure for every 10 m of depth).

Deep-sea organisms must be able to tolerate these *in situ* pressures. Indeed, many of these organisms will only grow at pressures exceeding atmospheric pressures. These are referred to as 'barophiles'; a term which was coined initially by Zobell and Johnson (1949). Certainly, studying barophiles is a highly specialised and imaginative branch of science.

It may seem surprising that the vast majority of the waters of the deep sea and the upper layers of the underlying sediment are oxygenated, with the amount of dissolved oxygen usually being greater than 50% of saturation levels for air, i.e. approximately 4 mg/l. This apparently high level of oxygen is directly attributed to the extremely low rates of oxygen consumption by the deep-sea organisms.

7.1.1   *Hydrothermal vents*

At the bottom of the three oceans, i.e. Atlantic, Indian and Pacific, volcanoes add to the topography of the sea floor. Many are active, emitting hot lava into the aquatic environment. The freshly exuded lava contracts during cooling, and permits seawater to percolate downwards through a distance of several kilometres into the newly formed crust (Jannasch and Taylor, 1984). Here, in the presence of high pressures and temperatures (>350 °C), the seawater reacts with the underlying basalt rock to form a highly reduced and acidic liquid, known as 'hydrothermal fluid'. This is enriched with hydrogen, hydrogen sulphide and metals, notably carbon, copper, iron, calcium, magnesium and manganese. Now, one of two possibilities occur. The hydrothermal fluid may rise and make contact with oxygen when mixing with the cooler seawater in the porous sub-surface lava. The result is the emergence of oxygenated (or non-oxygenated) *warm vents* at temperatures of 8–23 °C, with discharge rates of 1–2 cm/s. Alternatively, the hydrothermal fluid may discharge into the bottom waters without any prior mixing, creating oxic *hot vents* with temperatures of approximately 350 °C and discharge rates of 1–2 m/s (Jannasch and Mottl, 1985). The hot hydrothermal fluid, which contains particulates and hydrogen sulphide concentrations of 0.5–10 mM, mixes with cold seawater upon discharge from the vent. Thus, there is concomitant rapid and dense precipitation of grey-black polymetasulphide particles, which generate smoke-like plumes. These are called *'black smokers'*.

7.2   Experimental approach to the study of deep-sea microbiology

It cannot be said that deep-sea microbiology is short of ingenuity in experimental design, with techniques ranging from the very simple to complex. At one extreme, samples may be obtained in bags, grabs and corers from lines descending from surface ships. Here, there is no regard

to the problems of pressure and temperature differences between bottom
and surface waters. Upon recovery, the samples may be diluted and inoculated on to nutrient-limited media, e.g. seawater gelled with agar, or nutrient-rich media, e.g. marine 2216E agar (Difco), with incubation at
ambient (surface) temperatures for varying periods. The validity of such
data to explaining deep-sea microbiology is highly suspect. Some workers
re-pressurise already decompressed samples, thus highlighting the considerable degree of (baro-) tolerance of any micro-organisms which are capable
of surviving the experimental procedures. With increasing financial
resources, deep-sea submersibles, such as the Alvin with an operational
depth of 4000 m may be hired for the collection of samples. However, if
these are returned to the surface for examination at atmospheric pressure,
the extra cost of the operation is difficult to justify. One may therefore
question how deep-sea micro-organisms may be studied at *in situ* pressures
and temperatures. There are two possibilities, namely that samples be collected and returned to the surface at *in situ* pressures and temperatures,
or experiments be conducted on the sea floor. Both of these possibilities
are surrounded with endless problems, notably concerning the design of
appropriate equipment.

By the means of sophisticated pressure vessels, metabolic activity has
been shown to occur in the deep sea. The extent of metabolism reflects
the nature of the substrate used. In one set of experiments discussed by
Jannasch and Taylor (1984), the rate of metabolism of acetate was less
affected by pressure than glutamate metabolism. It is perhaps ironic that
upon reduction to atmospheric pressure, metabolism in deep-sea samples
invariably increases (Jannasch and Wirsen, 1982). Concerning the effects
of substrate on the rate of metabolism, it may be argued that different
substrates encourage the development of distinct types of micro-organisms
with various metabolic rates. It must be emphasised that this type of exper-
iment inevitably does not consider the nature of the microflora, solely the
ability to utilise substrates. Another explanation is that there are varying
responses to pressure by transport of different substrates and the subsequent
metabolic pathways. Only further elaborate experimentation will determine
which of these explanations is more appropriate to the deep sea.

In another set of experiments with a deep-sea sampler, Tabor *et al*. (1981a)
introduced glutamate at 50 μg/l into pressurised samples collected from a
variety of deep-sea locations. The results demonstrated that metabolic
activity by micro-organisms stopped before 2% of the glutamate was
utilised. This reaction to pressure was even more exaggerated than in the
experiments of Jannasch and Wirsen (1982), who determined that
metabolism of glutamate when dosed at 500–600 μg/l, ceased when 30%

of the substrate had been utilised.

Isolation of deep-sea micro-organisms has been achieved from pressurised samplers. For example, Dietz and Yayanos (1978) used a nutrient-rich silica gel shake tube medium to recover a spirillum-like bacterium from pressurised enrichments of a decomposing deep-sea amphipod, which was obtained from a depth of 5800 m. This organism grew best at pressures of 500 atm and 1000 atm, respectively, with a minimum generation time of 8–9 hours at 2–4 °C. The organism was truly psychrophilic, being sensitive to temperatures of >20 °C. Of particular interest, growth of the spirillum at 1 atm was <3% of growth at the optimum pressure (Yayanos *et al.*, 1979). Clearly, this organism was affected by decompression and would be severely stressed in any procedure which avoided the use of pressurised vessels. In subsequent work this team isolated an obligate barophile, which demonstrated an optimal generation time of 25 hours at the pressure of 690 atm, from an amphipod collected in the Marianas Trench at a depth of 10 500 m (Yayanos *et al.*, 1981). This organism failed to grow in pressures of less than 350 atm.

Agar plating techniques may be used in conjunction with pressurised vessels containing oxygen–helium mixtures to isolate bacteria. As a general rule, it appears that the optimum pressure for the growth of deep-sea barophiles corresponds closely to the *in situ* pressure from which the sample was obtained (Jannasch and Taylor, 1984). The use of a novel gelling agent, Gelrite, has been used in a nutrient-deficient medium to culture black smoker bacteria at 120 °C (Deming and Baross, 1986). Enrichment, e.g. with thiosulphate (1 mM), has enabled sulphur bacteria to be recovered from warm vents (Jannasch *et al.*, 1976). To date, most isolates of barophilic bacteria have been obtained from nutrient-rich niches (Yayanos *et al.*, 1979, 1982; Deming *et al.*, 1981), such as the gastro-intestinal tracts of deep-sea invertebrates. Generally such isolates only grow on nutrient-rich media.

In situ experiments on the sea floor have overcome some of the problems of recovery of samples and design of useful pressurised vessels. Experiments carried out on the sea floor include the degradation of materials, e.g. chitin and wood, and activity assessment using ^{14}C-labelled substrates. Again, data point to a reduction in rate of microbial metabolic activity with increasing depth. In their simplest form, deep-sea submersibles are used to position experimental material on the sea floor, with examination at periods of up to one year. Parallel experiments may also be carried out at atmospheric pressure (Jannasch and Wirsen, 1973).

Microbial activity in dissolved substrates may be examined by means of more elaborate equipment, such as the so-called 'syringe array'. Here, 6×200 ml syringes containing radio-labelled compounds were filled with seawater *in situ* through a common inlet nozzle by action of a mechanical arm on the deep-sea submersible. Incubation occurred *in situ* as well as in control sets at atmospheric pressure. Essentially this technique has been used to measure carbon dioxide fixation in the deep sea (Jannasch, 1984). The results have clearly demonstrated that organic compounds have a slower conversion in the deep sea than at atmospheric pressure (this assumes that temperatures are similar). Moreover, the *in situ* conversion rates were similar, regardless of whether the sample was collected at the surface or in deep water. To reinforce other experiments, this system also highlighted that there was a substrate-dependant relationship between increased pressure and metabolic rate.

Other *in situ* experiments have utilised independant vehicles, which are known as 'tripods' or 'bottom landers'. These descend freely to the sea floor for a variety of automated roles, returning to the surface after the timed release of an anchor (Isaacs, 1969). Such equipment has been used to measure microbial growth (Seki *et al.*, 1974), nitrate metabolism (Wada *et al.*, 1975), transformation of radio-labelled organic compounds in the upper layers of sediment, decomposition of compounds in the sediments, and metabolism of deep-sea amphipods and their intestinal microflora (Jannasch *et al.*, 1980; Wirsen and Jannasch, 1983).

Microscopy of deep-sea samples provides useful information about the size of microbial populations. Turner (1979) observed bacteria on the surface of faecal pellets, and concluded that many of the micro-organisms in the deep sea may be recent arrivals from surface waters. To some extent, this conclusion may be supported by the results of experiments which demonstrate an enhanced rate of metabolic activity with a reduction of pressure (e.g. Jannasch and Wirsen, 1982). Therefore, the supposition is that organisms are more active in shallower waters, with activity being retarded by increasing pressure and depth. Whereas this may be true of many deep-sea bacteria, there are some (as has already been stated) with growth optima at *in situ* pressures. Surely, these organisms must be true inhabitants of the deep-sea environment.

7.3   The microflora of the deep sea

7.3.1   *Quantitative aspects*

It should not be construed that microbial populations constitute a 'dilute-soup' of life in the deep sea. Instead, there are likely to be localised pockets of life, commensurate with available sources of nutrient. For example, the bacterial colonisation of faecal material has been observed (Turner, 1979). Moreover, 80% of the bacteria are found on particulates. In sediments, microbial activity is greatest in the uppermost strata being greater than in the water column and in the sediment layers only a short distance below the surface. From these data, it may be inferred that there is a deposition of micro-organisms from the water to the upper layers of sediment. For that matter, it has been determined that in the case of the respiration of acetate and hence its incorporation into particulate organic carbon, the ability decreased within the upper 7 cm thickness of deep-sea sediments (see Jannasch and Taylor, 1984). Nevertheless, using a deposition rate of 2.3 mm/1000 years for red clay (Arrhenius, 1952), it may be calculated that some of the organisms in deep-sea sediments were deposited over one million years ago (Morita, 1980). This begs the question about whether these organisms are active or dormant. As for numbers of organisms in sediment, 3.6×10^6 bacteria/g of wet sediment were reported by Schwartz *et al.* (1974). In this work, core samples were recovered from a depth of 4940 m and incubated at atmospheric pressure on a seawater; proteose peptone; yeast extract agar. Higher bacterial counts of between 2.1×10^8 and 4.7×10^8/g wet weight were found in sediment by Tabor *et al.* (1982). Of relevance was the correlation between counts obtained by epifluorescence microscopy and plating of dilute samples on marine 2216E agar (Difco) with incubation at atmospheric pressure. In a more refined set of experiments, Deming and Colwell (1985) used epifluorescence microscopy to determine the vertical distribution of bacteria in deep-sea sediments. Thus using cores collected at depths exceeding 4000 m, it was reported that bacterial populations in the top centimetre of sediment amounted to 4.65×10^8 bacteria/g dry weight. However, there was a doubling in numbers to 8.29×10^8 bacteria/g dry weight at a sediment depth of 3 cm followed by a progressive decline to 1.7×10^7 bacteria/g dry weight in a core sample collected at 15 cm from the surface of the sediment. Parallel results were obtained in a second core collected from a similar depth. Higher counts of approximately 3.07×10^{10} bacteria/g dry weight were recorded from faecal pellets. These counts were 9 to 72-fold higher than in the underlying surface sediment (Deming, 1985).

Deming (1985), using epifluorescence microscopy, estimated the bacterial populations of deep-sea water to be 1.44×10^8 bacteria/l. Interestingly, this count was approximately 20 times higher than results published by Carlucci and Williams (1978), but corresponds closely to the population density of 1.5×10^7–1.32×10^8/l reported by Tabor *et al.* (1981b). A sizeable proportion of these water-borne bacterial populations, i.e. between 0.5 and 77% of the total, comprised tiny cells which were capable of passing through the pores of a 0.45 μm porosity filter (Tabor *et al.*, 1981b). Seemingly, deep-sea bacteria may be extremely small in size.

As for the microbiology of deep-sea animals, Tabor *et al.* (1982) recorded populations of $<3 \times 10^4$/g and 2.1×10^9/g in the gastro-intestinal tracts of an amphipod and holothurian, respectively. Again, these data were obtained following incubation of diluted samples on a nutrient-rich medium, i.e. marine 2216E agar at atmospheric pressure.

7.3.2   *Quantitative aspects – hydrothermal vents*
As recently as 1977, populations of invertebrates were discovered in the vicinity of warm water hydrothermal vents, located at depths of approximately 2600 m (Ballard, 1977; Lonsdale, 1977). Numerically, there were greater populations of animals around the vents than at comparative depths in the water column away from the vents. Bacteria also occurred as cell suspensions in the warm vent waters, as microbial mats on surfaces in the immediate vicinity of the vent, and in symbiotic associations with invertebrates. The turbid waters of several of the warm water vents at the Galapagos Rift site contained populations of between 5×10^5 and 5×10^9 bacteria/ml as determined by epifluorescence microsocpy (Corliss *et al.*, 1979; Jannasch and Wirsen, 1979). Differences in population densities were explained by the irregular mixing of the warm vent waters with regular seawater. Nevertheless these counts were much higher than discussed previously for regular deep-sea water. The measurement of ATP provided results which were two to three-fold higher in warm vent water than in surface water at the same site. It was observed that the warm vent water contained flocculated particles, which upon microscopic examination were deduced to comprise bacteria (Jannasch and Wirsen, 1979). This was interpreted as meaning that bacteria produced mat-like growths within the warm water vent systems, which were subsequently transferred to the vented waters. In contrast, Jannasch (1985) did not detect significant bacterial numbers in hot vents as determined by epifluorescence microscopy or ATP measurements. Here there is a marked discrepancy with the findings of Baross and Deming (1983), who published bacterial counts of 4.7×10^5/ml in hot vent water at 304 °C. It has been suggested (Jannasch, 1985) that this discrepancy may reflect a pronounced variation in population sizes in different locations or point to the contamination of the samples in the rapidly cooling environment around the emission of the hot water. Only carefully executed experiments will clarify whether or not such hot waters contain integral microbial communities. To avoid undue speculation, it must be emphasised that water remains liquid at temperatures in excess of 100 °C when the hydrostatic pressure exceeds one atmosphere. For example, at a depth of 2600 m water does not boil below approximately 460 °C.

7.4 Taxonomy of deep-sea microbial populations

To date, all the available evidence points to deep-sea microbial populations as being comprised almost exclusively of Gram-negative bacteria. It is accepted that this situation undoubtedly reflects the state of current isolation methods but extensive microscopy has not provided evidence for the presence of sizeable populations of other microbes, i.e. Gram-positive bacteria, or fungi.

For isolations at atmospheric pressure, it would appear that bacteria of the type normally associated with surface waters predominate. Thus, *Aeromonas* (this is normally regarded as resident in fresh water), *Pseudomonas* – including *P. fluorescens* – and *Vibrio* have been recovered from the deep sea and identified according to an assortment of conventional and numerical taxonomy techniques (e.g. Quigley and Colwell, 1968; Schwartz *et al.*, 1974). As for the very small filterable bacteria reported by Tabor *et al.* (1981b), these were equated with *Alcaligenes, Flavobacterium, Pseudomonas* (numerically the dominant group) and *Vibrio*. Assuming that the samples were not contaminated with surface waters and that valid taxonomic conclusions have been made, it may be reasoned that these organisms may have originated from the upper levels of the water column. Isolations made at *in situ* pressures again reveal the presence of *Vibrio* (Deming, 1981) but also spirilla (Dietz and Yayanos, 1978). Thus, an impression may be gained that the deep sea does not contain interesting and novel micro-organisms. However, this notion is quickly dispelled upon consideration of the organisms associated with deep-sea hydrothermal vents; an aspect which has attracted a great deal of attention from established and novice taxonomists. A novel genus of copepod, i.e. *Isaacsicalanus*, has been recovered from warm vent waters (Fleminger, 1983). This component of the zooplankton was feeding on bacteria in the region of the interface between vent water and seawater. For that matter, exciting new groups of macro-organisms have been recognised in the vicinity of warm vents, including clams (*Calyptogena magnifica*), blue mussels (*Bathymodiolus* sp.) and the vestimentiferan tube worm (*Riftia pachyptila*).

Investigations of the bacterial components of warm vent waters and, in particular of the microbial mats occurring in and around the plumes, have focused on sulphur-oxidisers because of the widespread occurrence of reduced sulphur compounds in the environment. Indeed in one extensive study involving over 265 bacterial isolates recovered by enrichment on thiosulphate media, some acid-producing, obligate together with facultative chemolithotrophs were examined (Ruby *et al.*, 1981). Of the obligate chemolithotrophs, representatives of the genus *Thiomicrospira* predominated together with smaller numbers of *Thiobacillus* (Ruby and Jannasch, 1982). The *Thiomicrospira* were determined to be aerobic to micro-aerophilic, Gram-negative rods with mesophilic growth optima (Table 7.1), and a high tolerance to hydrogen sulphide, i.e. of up to 2 nM. Some isolates were equated with *Thiomicrospira pelophila* whereas others were quite different. At *in situ* pressures, i.e. 260 atmospheres, the organisms were only 75% as active as at atmospheric pressure. In that the 5s-rRNA sequence, G+C ratio of the DNA, optimal pH range for growth, and growth rate (2.6-fold higher than that of *T. pelophila* under equivalent conditions) of some isolates were different from those of the type species, *T. pelophila* (Stahl *et al.*, 1984), a new species was described, i.e. *T. crunogena* (Jannasch *et al.*, 1985). Unfortunately only scant information is available about the characteristics of these fascinating organisms (see Table 7.1). Further isolates have been recovered which have a wider optimum temperature range of growth, i.e. 26–33 °C. The generation (doubling) time has been reported as 1.25 hours (Jannasch and Nelson, 1984), which is the shortest of any obligate chemolithotrophic bacterial taxon.

All surfaces which are exposed to warm vents, including lava rocks, clam and mussel shells, worm-tubes and crab carapaces, are heavily colonised by bacteria. From studies with scanning and transmission electron micro-scopes, attempts have been made to identify the various microbial compo-nents (Jannasch and Wirsen, 1981). Indeed, genus names have been allo-cated to the various morphological forms. Although a degree of scepticism may be expressed about the wisdom of the allocation of names, especially

Table 7.1 *Characteristics of* Thiomicrospira *spp.*

Character	T. crunogena	T. pelophila
Gram-staining reaction	—	—
Cell size (μm)	0.4 × 1.5	0.2–0.3× 1–2
Motile by a single polar flagellum	+	+
Ability to oxidise sulphur	+	+
Chemolithoautotrophic	+	+
Aerobic growth	+	+
Optimum pH for growth	7.5 – 8.0	6.5– 7.5
Optimum temperature for growth (°C)	28 – 32	28– 30
Growth at 250 atmospheres of pressure	+	?[a]
G+C content of DNA (moles %)	41.8 – 42.6	43.9 – 44.0

[a]Question mark denotes don't know.
Data from Kuenen and Veldkamp (1972); Ruby and Jannasch (1982); Jannasch *et al.* (1985)

as some organisms have not been cultured, a viable alternative does not exist at present. Thus, cell filaments of 1–3 μm in width and up to 60μm in length, which could not be cultured, were considered to represent *Beggiatoa* and *Leucothrix* or *Thiothrix*. Such organisms appeared in mat samples as well as the vent water. *Hyphomonas–Hyphomicrobium*-like cells which possessed stalks were also observed. These organisms were isolated, and described as three new species of *Hyphomonas*, namely *H. hirschiana*, *H. jannaschiana* and *H. oceanitis* (Weiner *et al.*, 1985; Table 7.2). These organisms grew well at atmospheric pressure on marine 2216E agar. Gen-

Table 7.2 *Characteristics of* Hyphomonas *spp., recovered from the deep sea*

Characteristic	*Hyphomonas hirschiana*	*H. jannaschiana*	*H. oceanitis*
Gram-staining reaction	–	–	–
Rods (R) or cocci (C)	R	R	C
Presence of prosthecae	+	+	+
Presence of budding	+	+	+
Growth in the presence of air	+	+	+
Growth at atmospheric pressure	+	+	+
Growth in broth	Pellicle	Granular/ropy	Turbid
Generation time (h)	2–4	2–4	2–4
Optimum temperature of growth (°C)	25–31	37	20–30
Nutritional requirements:			
Tolerance to NaCl (% w/v)	2–15	2–15	1–7.5
Requirement for methionine	+	–	+
Requirement for biotin	–	+	+
Biochemical characteristics:			
Catalase production	+	+	+
Nitrate reduction	+	+	+
Oxidase production	+	+	+
Haemolysis (sheep blood)	γ	α	γ
Indole production	–	–	–
Hydrolysis of starch	–	–	–
Hydrolysis of gelatin	–	–	–
Production of arginine dihydrolase	–	–	–
Production of lysine decarboxylase	–	–	–
Production of ornithine decarboxylase	–	–	–
Production of deoxyribonuclease	–	–	–
Production of coagulase	–	–	–
Production of urease	–	–	–
G+C content of the DNA (moles %)	57	60	59

Data from Weiner *et al.* (1985)

erally they were fairly unreactive, being incapable of degrading blood, DNA, gelatin, starch or urea, or of producing arginine dihydrolase, coagulase, lysine or ornithine decarboxylase or indole. Despite the publication of these taxa in the *International Journal of Systematic Bacteriology*, the descriptions are not exhaustive, although it is conceded that they are *bona fide* representatives of *Hyphomonas*.

Electron microscopy also revealed the presence of heavily encrusted cyanobacteria which resembled *Calothrix*, anaerobic marine spirochaetes, methylotrophs, and unicellular cocci which were embedded in electron-dense compounds. These were identified by X-ray diffraction spectroscopy as ferromanganese deposits, correlating with the mineral, todorokite (Jannasch and Wirsen, 1981). Manganese-oxidisers were subsequently isolated (Ehrlich, 1983). Up to 15% of bacterial cells in the mats resembled Type I methylotrophs (Jannasch and Taylor, 1984), and so it is not surprising that an isolate of *Methylococcus* was recovered.

Thermophiles are particularly common in the vicinity of hydrothermal vents. The presence of these organisms was indicated by the observation that rates of tritiated-adenine incorporation into DNA and RNA were high when incubated at 90 °C (Karl, 1980). In particular, the rate of organic carbon production at 90 °C was equivalent to 19 µg C/g/h, which corresponded to a specific growth rate of 0.55/h (Jannasch and Taylor, 1984). Moreover in samples of hot vent water supplemented with inorganic salts and formate, the formation of methane and hydrogen was recorded after incubation at 100 °C ±2 ° (Baross *et al.*, 1982). Subsequently, thermophilic methanogens were recovered from the base of a 'hot smoker' chimney. A novel representative of *Methanococcus* was cultured, and constituted the first pure culture isolate of an anaerobic chemolithotroph from deep-sea hydrothermal vents. This organism was described as a new species of *Methanococcus*, as *M. jannaschii* (Jones *et al.*, 1983; Table 7.3). This anaerobic, irregular-shaped motile coccus grew optimally at 86 °C with a generation (doubling) time of 26 minutes. It is interesting to note that the polar membrane lipid was composed essentially (95%) of an unique macrocyclic glycerol diether (Comita and Gagosian, 1984) instead of the tetraethers, which are more common among the archaebacteria.

One isolate of an extremely thermophilic marine archaebacteria was recovered from a submarine hydrothermal vent at the East Pacific Rise. From its recovery with anaerobic culturing techniques incorporating 0.1% (w/v) yeast extract and 0.5% (w/v) peptone, the organism was recognised to be one of the archaebacteria because of the antibiotic resistance pattern, presence of glycoproteins in the cell envelope, and the possession of isopranyl ether lipids (Table 7.4). The organism was subsequently included in

Table 7.3 *Characteristics of* Methanococcus jannaschii

Character	Result
Colonial morphology:	1.5 mm in diameter, smooth, shiny, convex, circular and yellow after two days
Micromorphology:	Irregular-shaped coccus with osmotically fragile cell wall. Width $= 1.5\,\mu m$
Motility	+ (polar bundles of flagella)
Generation (doubling) time (min)	26 (at 85 °C)
pH range of growth	5.2–7.0
Autotrophic	+
Ability to produce methane	+ (from methylcoenzyme M, with hydrogen or formate as electron donors)
Obligate anaerobic growth	+
Temperature range of growth (°C)	50–86
G+C content of DNA (moles %)	31

Data from Jones *et al.* (1983)

Table 7.4 *Characteristics of* Staphylothermus marinus

Character	Result
Cell envelope composition	>4 polypeptides; absence of muramic acid.
Lipids	20% glycerol diethers, 40% dibiphytanyl diglycerol tetraethers; 40% unidentified ether lipid components
Micromorphology	Regular to slightly irregular-shaped Gram-negative cocci of 0.5–1.0 μm in diameter, occurring singly, in short chains, and in clumps of up to 100 cells. Giant cells, of $\leq 15\,\mu m$ in diameter, are formed in high concentrations of yeast extract.
Motility	—
Generation (doubling) time (min)	270
Growth in sodium chloride (% w/v)	1–3.5
pH range of growth	4.5–8.5
Heterotrophic	+ (growth occurs on peptone and yeast extract)
Requirement for elemental sulphur	+
Obligate anaerobic growth	+
Metabolic products	Acetate, carbon dioxide, hydrogen sulphide, isovalerate
Temperature range of growth (°C)	65–98 (optimally at 92)
G+C ratio of the DNA (moles%)	35
Antibiotic resistance (150 $\mu g/ml$)	Chloramphenicol, kanamycin, streptomycin, vancomycin

Data from Fiala *et al.* (1986)

a new genus, *Staphylothermus*, as *Staphylothermus marinus* (Fiala *et al.*, 1986). A second isolate was recovered from geothermally-heated beach sediments in Italy. The similarity between these two isolates was confirmed as a result of DNA–DNA hybridisation (>90% DNA homology).

7.5   Activity and role of deep-sea micro-organisms

In the majority of cases, away from hydrothermal vents, deep-sea micro-organisms must be effectively struggling for survival in a state of near-starvation. One of the most dramatic effects is that cells become much smaller, and may be quite capable of passing through the pores of a bacteriological filter (Tabor *et al.*, 1981b). Yet an obvious benefit of this near-starvation is an increase in barotolerance. Thus it may be argued that there is an ecological advantage in the deep sea to tolerance of starvation conditions and hence an increase in barotolerance. Considerable debate has centred on the precise effects of pressure on bacterial cells. Although not proven, it has been suggested that the synthesis of macro-molecules is more affected by pressure than is substrate respiration (Tabor *et al.*, 1981a,b). In a separate investigation, Baross *et al.* (1978) used barophiles recovered from lesions on tanner crabs (*Chionoecetes tanneri*), which live at depths of 500–5000 m. Using an isolate, designated TC-L1-10, the effects of pressure of up to 1000 atm at 3.5 °C were assessed on binding, transport, respiration and protein synthesis. Experiments measured the uptake of ^{14}C-glutamic acid at 2 hours. Effectively, the results demonstrated that there was a decrease (concomitant with increasing pressure) of substrate binding, transportation of the amino acid across the membranes and into the cell, and of protein synthesis. This decrease was associated with a concomitant rise in the level of carbon dioxide respired. It is noteworthy that Paul and Morita (1971) reported that enzymes involved in the dissimilation of amino acids to carbon dioxide were not as susceptible to the effects of pressure as was the mechanism of enzyme transport. Thus, it may be deduced from these experiments that the primary effect of pressure is on the cell membrane. Clearly, further experiments are necessary to clarify this situation.

The harshness of conditions in the deep sea is illustrated by the lengthy mean generation (doubling) times of cells. Although there are conflicting views as to the precise times involved, at least one study has reported a link between doubling times and depth. For example, Carlucci and Williams (1978) estimated *in situ* doubling times to be 145 h at a depth of 1500 m, extending to 210 h for a depth of 5550 m. One criticism of the data is that they were derived from pure culture study and may therefore not reflect the circumstances in the natural environment. Indeed, a much shorter generation time of 4.3–4.5 h at depths of 4120–4715 m was suggested by

Deming (1985). Yet, the slow-growing nature of deep-sea bacteria has been highlighted in experiments aimed at enriching for sulphate-reducers at a pressure of 700 atm. Here, sulphate reduction could not be detected with incubation periods of less than 10 months at 5 °C (see Morita, 1979). Ironically, control experiments carried out at atmospheric pressure were negative for sulphate reduction even after two years.

Even when favourable nutrient conditions exist, microbial activity in the deep sea is exceptionally slow. This was highlighted by the fortuitous events associated with the accidental sinking of the deep-sea submersible, Alvin, in 1968. The hatch was open and, therefore, the vessel was subjected to *in situ* conditions until recovered after 10 months, from a depth of 1540 m. It so happened that a consignment of organic material (sandwiches) contained within a tightly closed plastic box (sandwich box), was found in a remarkable state of preservation after the submersion (Jannasch *et al.*, 1971). Yet this material deteriorated rapidly at atmospheric pressure in a refrigerator. So at similar temperatures but vastly different pressures, there was a marked contrast in biodegradation of the organic material. This was supported by subsequent controlled experiments. However, in hydrothermal vents, generation times of only 26 minutes have been recorded for some bacteria (Jones *et al.*, 1983), which indicates a comparatively high metabolic activity. Nevertheless, it has been demonstrated that growth of bacteria in warm vents ceases if the temperature drops to that of ambient seawater, i.e. 2.1 °C (Tuttle *et al.*, 1983). By estimating CO_2 incorporation, it was concluded by Jannasch (1984) that vent microbial populations were barotolerant and mesophilic. These deductions resulted from experimental data which revealed *in situ* levels of CO_2 incorporation of 1 nmol/l/day at 3 °C outside the plumes. Similar levels occurred at atmospheric pressure and a temperature of 3 °C. Yet at 25 °C, CO_2 incorporation was determined to be as much as 100 nmol/l/day at atmospheric pressure.

In an ingenious set of experiments, Barber (1968) concentrated the dissolved organic component of surface and deep-sea water, and incubated it with the indigenous microflora for two months. The results demonstrated that, in contrast to surface samples, there was no change in the quantity of dissolved organic compounds in deep-sea water although viable bacteria were present. Clearly, the metabolic activity of these organisms must have been incredibly slow, perhaps corresponding to the notion of dormancy.

One of the important roles of the deep-sea bacteria is that of serving as a primary food source for aquatic invertebrates. Certainly it seems likely that bacteria serve as the primary source of carbon, nitrogen, some amino acids and energy for invertebrates (detritivores). The long-chain polyunsaturated fatty acids, which were thought to be absent from prokaryotes but have been identified in deep-sea bacteria, are also probably important in the food chain (De Long and Yayanos, 1986). In this case, any activity leading to the demise of the deep-sea micro-organisms would have drastic consequences for the survival of all life forms.

7.5.1   *Chemosynthetic activity at deep-sea hydrothermal vents*

In hydrothermal vents, non-phototrophic, chemoautotrophic, bacteria comprise the primary producers of organic carbon, using reduced inorganic compounds, notably hydrogen sulphide, thiosulphate, ammonia, nitrite, hydrogen, carbon monoxide and iron, in the hydrothermal fluid as a source of geothermal (chemical) energy in a process known as chemosynthesis (Jannasch and Wirsen, 1979; Karl *et al.*, 1980). However, it should be emphasised that the concept of primary productivity by chemosynthesis has been questioned (see Karl, 1987). Nevertheless, the process (if it indeed occurs) is thought to be aerobic, requiring free oxygen. Moreover, it has been reported that chemosynthesis occurs within three distinct areas of the hydrothermal vents, namely in microbial suspensions emitted with the warm vent water, in the microbial mats and in symbiotic associations between invertebrates and chemosynthetic bacteria. These newly discovered types of symbioses (Cavanaugh, 1985) involve the invertebrates, i.e. clam, blue mussel (*Bathymodiolus* sp.) and vestimentiferan tube worm (*Riftia pachyptila*), found in unusual predominance around the vents, and bacteria. Both enzymic and histological studies have pointed to the major transfer of chemosynthetically-produced organic carbon to the animals. In the case of the vestimentiferan tube worm, *Riftia pachyptila*, detailed electron microscopy has revealed the presence of short rods and large (3.5 μm in diameter) spherical cells at concentrations of up to 10^9/ml (Jannasch, 1984). A morphologically similar organism was determined to comprise a hydrogen-oxidiser with a G+C ratio of the DNA of approximately 60 moles %, which was considered to resemble *Thiobacillus*. This organism was not stimulated by reduced sulphur compounds.

8   Benefits and malefits of marine micro-organisms

It is all very well to provide information about the size, nature and activity of marine micro-organisms; however, laymen are often only interested in the benefit or harm of such microbes to human society. One of the beneficial applications is in the realm of biotechnology, which will be considered separately (see Chapter 9). Other benefits, and conversely, malefits are discussed below.

8.1   Benefits

8.1.1   *Biodegradation of pollutants*

Marine micro-organisms have a marked ability to degrade complex molecules, including naphthalene (Raymond, 1974, Voronin *et al.*, 1977), pesticides (Gibson and Brown, 1975), wood pulp (Vance *et al.*, 1979) and petroleum. Certainly the last-mentioned ability has been the focus of great attention in accidents where the discharge of hydrocarbons from damaged ships, such as the Amoco Cadiz and the Torrey Canyon, has occurred.

Petroleum is a complex mixture of many thousands of compounds including gases and residues which boil at temperatures of 400 °C. The components include *n*-alkanes, which are the most readily degraded, iso-alkanes, cyclo-alkanes, aromatics, naphthenoaromatics, heterocyclics (resins) and asphaltenes. It has been estimated that the annual input of hydrocarbons into the marine environment falls somewhere in the range of 1.9×10^6 tonnes (Beastall, 1977) to 6.11×10^6 tonnes (Johnston, 1980). In UK waters, the annual input of oil has been reported as 40–100 kt (Whittle *et al.*, 1982). Most of this input occurs as a result of losses during transportation and from drilling operations, including disposal of cuttings (Johnston, 1980).

It is well established that micro-organisms which utilise, oxidise and degrade hydrocarbons occur in the marine environment. Zobell (1969) noted that such organisms occurred in water and sediment collected between latitudes of 40 °S and 60 °N. In unpolluted conditions, hydrocarbon-degrading (referred to as 'hydrocarbonoclastic') organisms comprise less than 2% of the total microflora (Hughes and McKenzie, 1975). Yet in the presence of such pollutants, the proportion of hydrocarbon-degrading organisms is substantially increased; and with chronic discharge of petroleum, microbes with degradative ability may predominate (Atlas, 1981). Zobell and Prokup (1966) recovered 10^2–10^8 colony-forming units/g of sediment with the ability to degrade lightweight paraffin oil in Barataria Bay, USA. Using model petroleum substrate, Walker *et al.* (1976) reported a progressive increase in the number of degraders with distance from shore. Thus, at sampling sites 250 m, 50 km and 375 km from shore, the number of hydrocarbon-utilising organisms was 1.5×10^2–2.3×10^2/ml, 2.8×10^2–3.3×10^3/ml and 4.0×10^4/ml, respectively.

The hydrocarbonoclastic organisms encompass a fairly diverse range of taxa. Zobell (1973) considered that 70 microbial genera, including 28 bacterial genera, 30 genera of filamentous fungi and 12 genera of yeasts, possessed representatives which were capable of attacking hydrocarbons. Of these organisms, the bacteria were regarded as the most important in terms of the range of hydrocarbons which may be attacked (Soli, 1973). Relevant taxa included *Acinetobacter*, actinomycetes (mycelial), *Arthrobacter, Brevibacterium, Corynebacterium, Flavobacterium, Klebsiella aerogenes, Micrococcus, Mycobacterium, Nocardia, Pseudomonas, Sphaerotilus natans* and *Vibrio*. The fungal-degraders included representatives of *Candida, Cladosporium, Rhodotorula, Sporobolomyces, Torulopsis* and *Trichosporium*. Seemingly, the combined (possibly synergistic) activity of these organisms ensures that the seas are not permanently covered with an oily layer (Atlas, 1977). Nevertheless it is conceded that the degradation of hydrocarbons is a slow process.

Undoubtedly, the majority of work concerning petroleum degradation has employed *in vitro* methods. Consequently, there will be a perpetual problem of relating the results of laboratory experiments to conditions in the natural environment. In field studies, enrichment techniques are most commonly employed, whereas expensive equipment may dominate in laboratory experiments. Work has encompassed the toxicity effect of hydrocarbons on selected indicator species (Dicks, 1982; Dicks and Hartley, 1982), gas–liquid chromatography (Soli and Bens, 1972) and mass-spectroscopy (Walker *et al.*, 1976). These more sophisticated approaches measure microbial growth or some aspect of metabolic activity. Enrichment techniques using single hydrocarbons or model petroleum do not necessarily generate results reflective of events in the polluted marine environment. With radio-labelled compounds there is the added problem of disposal, however, an advantage of this method is that an assurance will be gained that the labelled substrate is actually attacked. Pure culture work in the laboratory may be misleading insofar as co-metabolism may be important in the degradation of hydrocarbons within the natural environment. Nevertheless, it would appear that degradation generally proceeds by the introduction of oxygen into the molecule. In the case of bacterial attack, the fission of ringed moieties requires dihydroxylation (Rosenberg and Gutnick, 1981). Factors affecting the rate of degradation include availability of oxygen, a supply of suitable inorganic nutrients, pH and temperature.

Hydrocarbon degradation is a strictly aerobic affair. Consequently an adequate supply of oxygen is necessary, and it seems unlikely that anaerobes exert a significant influence on degradative processes. Of the essential inorganic nutrients, a role for nitrogen as nitrate (NO_3^-) and phosphorus as phosphate (PO_4^{3-}) has been demonstrated. Thus, Beastall (1977) calculated that 6–60 mg of nitrogen and 1–10 mg of phosphorus were necessary to degrade 1 g of oil. Moreover, it was estimated that maximal degradation occurred at warmer temperatures of 25–37 °C (Beastall, 1977). Consequently in the colder temperatures of the open sea, a low rate of biodegradation occurs. Some laboratory experiments have provided useful pointers to the rate of biodegradation. For example, Pritchard and Starr (1973) calculated that 16.2–57.1 μg of octane could be degraded in an hour by using continuous culture methods. Using sand columns, Hughes and McKenzie (1975) determined that 78–84 g of oil were degraded within seven days, although this rate reduced to 2.9–3.3 g/week as the proportion of simpler (readily degradable) compounds declined. Ramibeloarisoa *et al.* (1984) used mixed bacterial populations, which were derived from surface foams of chronically polluted areas in the Mediterranean Sea. Essentially, the data revealed that these bacterial populations degraded crude oil in seawater which was supplied with iron, nitrogen and phosphorus. In fact, 81% of the oil was degraded when the conditions included a temperature of 30 °C, pH 8 and full oxygenation. After 12 days, the amounts of saturated, aromatic and polar compounds, and asphaltenes degraded corresponded to 92%, 83%, 63% and 48% of the totals, respectively. Furthermore, it was apparent that bacterial growth on hydrocarbons induced the production of surfactants, which emulsified the substrates. Consequently, it was concluded that biosurfactants are important in the elimination of hydrocarbons from the marine environment.

So, the literature points to the ability of micro-organisms to contribute to the degradation of hydrocarbons in the marine environment. Therefore, it is pertinent to enquire whether such research may aid the polluter in cleansing the environment. Perhaps the notion of developing a genetically-engineered 'superbug' (Horowitz and Atlas, 1980), which could be sprayed on to oil slicks, is not too far-fetched. Under carefully controlled conditions, this could open up prospects for the production of single-cell protein as a valuable by-product. However, at present, such an organism does not exist.

8.1.2   *Role of micro-organisms in the settling of invertebrate larvae*

It was initially suggested by Zobell and Allen (1935) that marine bacteria are involved in the settlement of invertebrate larvae to solid substrates. Since then, it has become increasingly obvious that micro-organisms exert a tremendous influence on the ability of larvae to settle as a prelude to metamorphosis (Scheltema, 1974), and for other macro-organisms, such as barnacles and algae, to attach to substrates (Crisp, 1974). Furthermore it has been argued convincingly that there is the potential to exploit this beneficial interaction between microbes and invertebrates for biotechnology (Bonar *et al.*, 1986). The mechanism of this preferential settlement appears to be related to the production of microbial compounds. Thus, Neumann (1979) reported the ability of *Vibrio* ultrafiltrates, in the size range of 10^3–10^4 daltons molecular weight, to induce the settlement of larvae of the scyphozoan cnidarian *Cassiopea andromeda*. Kirchman *et al.* (1982a,b) considered that a polysaccharide or glycoprotein on the surface of biofilms, produced by *Pseudomonas marina*, induced settlement and metamorphosis of the polychaete, *Janua* (*Dexiospira*) *brasiliensis*. In this example, the stimulus for the occurrence of metamorphosis was thought likely to involve the binding of larval lectins with the bacterial exo-polymer. Similarly, bacterial polysaccharides have been implicated in the settlement of oyster larvae (Weiner and Colwell, 1982). Another settlement-inducer is the possible stimulation of larval (active) cation transport, notably Na^+/K^+–ATPase (Müller and Buchal, 1973). However, it remains to be seen whether or not micro-organisms or their exo-polymers will be produced and used commercially for the induced settlement of economically important marine invertebrates.

8.1.3   *Microbial involvement with the formation of manganese nodules*

It has been recognised for some considerable period that manganese nodules are scattered over the sea-bed. Indeed, there are grandiose commercial schemes to collect these nodules. However, there is considerable controversy over the precise origin of these deposits. A topical hypothesis implicates bacteria in the formation of the nodules. Certainly it has been established that many marine bacteria, such as representatives of *Achromobacter, Aeromonas, Arthrobacter, Bacillus, Brevibacterium, Flavobacterium, Micrococcus, Oceanospirillum, Pseudomonas, Siderocapsa* and *Vibrio*, have the ability to precipitate manganese as an insoluble stable oxide, i.e. MnO_2 (see Nealson and Tebo, 1980). Moreover, it has been established that some marine bacteria possess extracellular capsules containing the precipitate (Cowen and Silver, 1984). Such bacteria, which have been mostly described as Gram-positive cocci of 0.3–0.5 μm in diameter (including capsular material, the diameter increases to 1–2 μm) are found with increasing frequency below depths of 100 m. Indeed, the frequency of these capsular organisms increases with depth. Therefore it is tempting to equate these organisms to the presence of manganese on the sea floor.

8.1.4   *Fermented food products*

In the industrialised nations of the Western World, microbial interactions with food derived from the sea are generally regarded as detrimental to both the product and the consumer. Yet in Eastern countries, many fermented foods result from microbial activity. For example, a product named Shoyu results from microbial activity on brown algae. Other algal products have been derived from the activity of *Alternaria, Aspergillus* and *Penicillium*. Fermented fish dishes result from the effects of moulds and/or clostridia. A squid meat product, known as Ika-Shiokara, involves yeasts in the ripening process. In the presence of 20–30% (w/v) sodium chloride, the dominant yeasts comprise *Rhodotorula mucilaginosa* and *R. minuta* var. *texensis*, with smaller numbers of *Candida, Cryptococcus, Debaryomyces* and *Sporobolomyces*. However, with only 10% (w/v) sodium chloride, the dominant yeast is initially *R. mucilaginosa* with a succession to *D. kloeckeri* after one week into the ripening process. In this concentration of salt, the other yeasts include *Candida, Cryptococcus, Torulopsis* and *Trichosporon* (Mori *et al.*, 1977).

Finally, it may be argued that the presence of certain organisms may be used as a valuable indicator of polluted conditions.

8.2   Malefits

8.2.1   *Biodeterioration/biofouling of objects in the sea*

Marine micro-organisms have also gained notoriety for their role in biodeterioration of useful objects, such as wooden dockyard pilings, cotton fishing nets and ropes. Here there is more likelihood that natural substances will be attacked rather than synthetic materials, such as nylon. Biofouling of ship undersurfaces may have dire economic significance, insofar as fouled ships move more slowly in the water than their unfouled counterparts. Copper-based paints, which release 10–20 μg of copper/ cm^2/ day, and/or organotin compounds are invaluable in the prevention of biofouling.

The pattern of events leading to biofouling appears to involve initial attachment by simple Gram-negative bacteria, followed by the arrival of stalked bacteria (Crisp, 1974), then diatoms (Colwell *et al.*, 1980) – notably *Bellerochea, Biddulphia, Chaetoceros* and *Melosira* (Wood, 1967) – and protozoa. Thus, there is an initial low diversity of organisms followed by a greater diversity.

Colwell *et al.* (1980) studied the processes leading to biofouling and thence biodegradation of wooden dockyard pilings by the isopod *Limnoria tripunctata* in a tropical marine harbour. Essentially, the data pointed to the establishment of biofouling communities, i.e. mat-like coverings of floc-forming bacteria, within two days, even on pilings which were impregnated with 40% creosote–naphthalene. However, the rate and extent of fouling was greater on untreated than treated wood. Ultimately, the pilings were destroyed at the air–water interface (Figs. 8.1, 8.2).

8.2.2   *Mobilisation of heavy metals*
The ability of micro-organisms to concentrate heavy metals, e.g. mercury (Colwell *et al.*, 1975), may have dire consequences for macro-organisms insofar as the potential toxic material may enter food webs. For example, bacteria (particularly *Pseudomonas*) may take up and store mercury, possibly as a result of plasmid-mediated activity. These bacteria may in turn be concentrated in filter feeders, e.g. oysters, or be consumed by eukaryotes, particularly ciliates, notably *Keronopsis pulchra* (Colwell *et al.*, 1975) and *Uronema nigrificans* (Berk and Colwell, 1981). In turn, the ciliates are food for copepods, such as *Eurytemora affinis* (Berk and Colwell, 1981), which may be consumed by fish. Thus, from the initial uptake by bacteria, the toxic metals may be quickly transferred along the food chain.

8.2.3   *A reservoir for human pathogens*
Unfortunately there may be profound consequences to public health of the presence of certain (human) pathogens in the marine environment, e.g. *Vibrio cholerae* (Müller, 1977). Whereas the source of some of these organisms may be traced to sewage discharges, other pathogens appear to be normal residents, especially of coastal waters. Faecal organisms, such as *Escherichia coli*, may be removed by predatory amoebae, e.g. *Vexillifera*, and fruiting myxobacters of the genus *Polyangium* (Roper and Marshall, 1978). However, *Vibrio parahaemolyticus* is common in coastal and estuarine waters, sediment and invertebrates, particularly during the summer (El-Sahn *et al.*, 1982). This organism, together with *Bacillus cereus, Clostridium perfringens* and occasionally *Salmonella*, may accumulate in bottom-dwelling filter feeders, e.g. mussels and oysters (Van den

Fig. 8.1 Biodeterioration of a wooden piling in a tropical harbour.

Fig. 8.2 Decayed wooden piling, as a result of microbial activity.

Broek *et al.*, 1979; Thi-Son and Fleet, 1980). If consumed raw by human beings, they may cause food poisoning. Thus, the need for sound depuration practices is obvious.

8.2.4   *Spoilage and food-poisoning micro-organisms in fish*

It is controversial whether or not fish muscle is normally sterile in whole unfilleted fish (Bissett, 1948; Shewan, 1971). However, from capture, the fish may be subjected to contamination during storage and filleting. On fishing ships, hygiene may be of questionable standard. For example, the microflora of drainage water from ice has been determined to comprise as many as 2.1×10^7–2.2×10^9 bacteria/ml and 6.3×10^3–7.2×10^4 filamentous fungi and yeasts/ml (Chen and Chai, 1982). Here the dominant taxa included representatives of *Acinetobacter, Alcaligenes, Bacillus, Corynebacterium, Flavobacterium, Micrococcus, Moraxella* and *Vibrio*. By the time they reach shore, the whole fish may have small microbial populations in the muscle, e.g. 2×10^3 bacteria/ml (Alexander and Austin, 1986). Undoubtedly, the cleanliness of the subsequent filleting procedure, i.e. condition of the knives, cutting boards and wash water, and storage facilities influences the size of the microbial populations on fillets. Thus, Alexander and Austin (1986) recovered 9.2×10^5 bacteria/g from the muscle of freshly filleted haddock. Many of these organisms could contribute to spoilage. In particular, bad odours result from the activity of *Alteromonas* and *Pseudomonas*; hypoxanthine is produced from inosine-5-monophosphate by *Alteromonas, Photobacterium* and *Pseudomonas putrefaciens* and trimethylamine is produced from trimethylamine oxide by moraxella-like bacteria, *Photobacterium* and *Pseudomonas putrefaciens* (Van Spreekens, 1977). At this stage the fish would be unsaleable.

Histamine-forming bacteria, which have relevance in allergic-type food-poisoning, have been recovered from fresh and spoiled fish (Okuzumi *et al.*, 1984). Such organisms, which may reach populations of 10^6–10^8/g on spoiled fish, have been identified as *Aeromonas, Citrobacter, Hafnia alvei, Proteus morganii, Proteus vulgaris* and *Vibrio*. Levels of >100 mg of histamine/100 g of fish could result in clinical illness.

9   Biotechnology

Microbiologically, the years leading from the late 1970s will probably be recorded in history books as the era of biotechnology. The justification for funding seemed better if the likely outcome of fundamental basic research was perceived in terms of potential commercial exploitation of the results. This was not a new concept insofar as brewers have essentially been carrying out biotechnological processes with yeasts for many hundreds of years. None the less the term biotechnology has captured public attention, and research programmes have been re-directed accordingly. However, the role of biologists in technological processes appears rosy, with the emphasis on antibiotics, chemicals, enzymes, polymers, vitamins and single cell protein. Undoubtedly, soil has contributed the overwhelming majority of commercially useful strains of micro-organisms to date. However, it is pertinent to enquire whether or not the marine environment should be considered as worthy of attention by biotechnologists. The answer to such a question must be affirmative, insofar as marine micro-organisms comprise a comparatively untapped reservoir of commercially valuable compounds. It is apt to recall previous discussion which highlighted the marked ability of some marine bacteria to produce and secrete polymers and enzymes. Indeed, it is the author's personal experience that some marine bacteria are potent producers of DN'ases, lipases, alginases and proteases. Where the ability is present, enzymes are often secreted in plentiful supply. As a topical example, a single bacterial colony may clear DNA, casein or gelatin from the entire media contained in Petri dishes. This is no mean feat compared to soil organisms. Therefore it must be argued that the marine environment deserves rightful emphasis in the sphere of biotechnological influence. Such interest may focus in a number of areas, including production of biochemical compounds, enzymes, single cell protein and pharmaceutical compounds.

9.1   Pharmaceutical compounds

9.1.1   Antibiotics

The scientific literature abounds with references to antibiotic production among marine organisms, but the pharmaceutical industry has been slow in exploiting the data. Indeed, the earliest indication of potentially inhibitory marine bacteria is derived from the work of De Giaxa (1889), who reported the presence in seawater of bacteria, which were antagonistic to the causal agents of anthrax and cholera. Then after a lull of nearly 60 years, Rosenfeld and Zobell (1947) described the production of antibiotics by marine bacteria. Interestingly, in this study the majority of the antibiotic-producers were equated with *Bacillus* and *Micrococcus*. These are usually regarded as terrestrial organisms rather than representatives of the true marine microflora. Further work by Krassil'nikova (1961) and Buck *et al.* (1962) confirmed antibiosis among marine bacteria, with the latter study pointing to inhibitory effects against yeasts. Then a proliferation of research resulted in numerous publications, starting in 1966. Of these, the work of Burkholder *et al.* (1966) is relevant insofar as characterisation of the inhibitory compound revealed a novel chemical structure. This was confirmed by Lovell (1966) as 2,3,4-tribromo-5(1'-hydroxy-2',4'-dibromophenyl)-pyrrole. The compound was recovered from bacteria, which had been isolated from *Thalassia* sp. (turtle grass) collected in the vicinity of Puerto Rico. Isolation was relatively straightforward, using a pour plate technique with nutrient-rich media (Table 9.1). The sample of *Thalassia* (weighing approximately 1 g) was shaken in a 99 ml aliquot of sterile seawater and 1 ml volumes incorporated in pour plates with incubation at room temperature overnight. Examination of the plates revealed the presence of yellowish colonies, some of which appeared to cause a darkening in the medium. Burkholder and co-workers subsequently studied two isolates, referred to as No. 287 and No. 396. These were reported as Gram-negative, antibiotic-producing, motile rods, which grew at 20–25 °C and required seawater. Gelatin, starch and urea, but not agar, were attacked. H_2S was produced by one organism, but not so catalase or indole. The methyl red and nitrate reduction tests and the Voges Proskauer reaction were negative. Resistance was noted to the vibriostatic agent, O/129. A range of carbon compounds was utilised aerobically, including fructose, glucose, maltose, mannose, sodium citrate, sucrose and trehalose, but not adonitol, arabinose, dulcitol, glycerol, inositol, inulin, lactose, mannitol, rhamnose, salicin, sorbitol or xylose. From these characteristics and despite a negative reaction for the catalase test, the organisms were considered to resemble *Pseudomonas*, for which the name of *Pseudomonas bromoutilis* was proposed to accommodate isolate No. 396 on account of the bromopyrrole production. Alas, this taxon is not included in the *Approved Lists of Bacterial Names* (Skerman *et al.*, 1980) or their supplements, and therefore lacks taxonomic validity. Nevertheless, the organism, recovered using non-selective isolation techniques, demonstrated pronounced inhibitory activity. Moreover superior antibiotic production occurred on solid media rather than liquid shake cultures. Thus, the inhibitory compounds were recovered by using a modification of the isolation medium (Table 9.2). This was inoculated, and then incubated at 30 °C for three days after which the cells were scraped off prior to freezing at −15 °C. Thence, the inhibitory compound was extracted using methanol, ethyl ether and chloroform (Fig. 9.1). After fractionation, the antibiotic was collected, dissolved in chloroform, and cooled to −10 °C whereupon crystallisation occurred. Weakly-green long needles of antibiotic developed which decomposed at 135–155 °C. These were very soluble in acetone, ethyl ether and ethyl acetate, and moderately soluble in chloroform and methanol but were insoluble in water. The antibiotic was inhibitory to Gram-positive bacteria, notably *Staphylococcus aureus, Streptococcus pneumoniae* and *Streptococcus pyogenes* at concentrations of only 0.0063 μg/ml, and to *Mycobacterium tuberculosis* at 0.2 μg/ml. In contrast, the compound was ineffective against Gram-negative bacteria.

Table 9.1 *Medium for the recovery of antibiotic-producing marine bacteria*

Ingredient	% (w/v) Composition
N–Z case	0.2
Soytone	0.1
Yeast extract	0.1
Agar (Difco)	2.0
Experimental conditions; prepared in seawater	

After Burkholder *et al.* (1966).

Table 9.2 *Medium for the recovery of inhibitory compounds from* 'Pseudomonas bromoutilis'

Ingredient	% (w/v) Composition
N–Z case	0.2
Soytone	0.2
Yeast extract	0.1
Dextrose	0.2
Vitamin B_{12}	0.000 0001
Agar (Difco)	2.0
Experimental conditions; prepared in seawater; sterilised at 121 °C/20 min.	

After Burkholder *et al.* (1966).

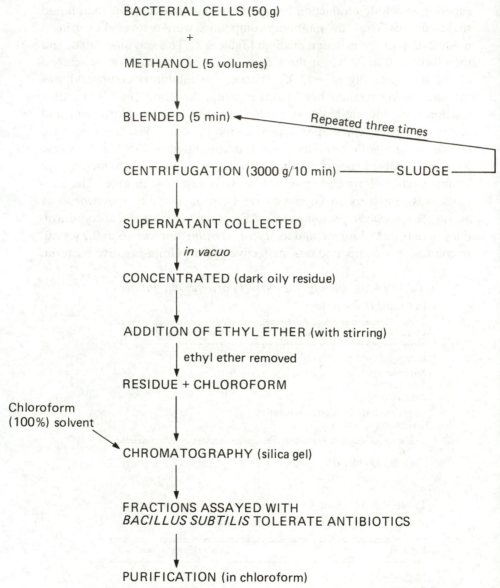

Fig. 9.1 Method for the extraction of antibiotics from '*Pseudomonas bromoutilis*' (after Burkholder *et al.*, 1966).

So a marine organism produced a compound which was effective against Gram-positive pathogens. This compound was not exploited further, which in retrospect seems to be a pity because of the current widespread problems of antibiotic resistance, especially among clinical strains of *Staphylococcus aureus*.

Within a decade of this work, another inhibitory bromopyrrole was recovered from marine bacteria. Using non-selective isolation techniques, Anderson *et al.* (1974) recovered an antibiotic-producing, purple-pigmented *Chromobacterium*, designated strain 1-L-33, from seawater in the North Pacific. The organism was described as motile, Gram-negative rods, which produced oxidase but not catalase or indole, hydrolysed starch, and gave negative responses to the methyl red test and Voges Proskauer reaction. From these traits, the organism was equated with *Chromobacterium*, and, in particular, considered to resemble *C. marinum*. In fact because of the production of bromopyrroles, the organism was linked to *P. bromoutilis*. However from the published descriptions of the two organisms, it is not possible to firmly agree (or even disagree) with this suggestion.

The purple-pigmented *Chromobacterium* was easily maintained with regular sub-culturing in laboratory conditions (Table 9.3). Mass culturing was achieved by incubation with aeration at 25 °C/14 days in broth (Table 9.4) or as colonies on solid medium (Table 9.5) after incubation at 25 °C/4 days. Essentially, the inhibitory products were extracted in ethyl acetate (Fig. 9.2), and equated with 4-hydroxybenzaldehyde, tetrabromopyrrole (Fig. 9.3), which was unstable and light sensitive, hexabromo-2,2′-bipyrrole, and

Table 9.3 *Medium for the routine maintenance of an antibiotic-producing purple marine* Chromobacterium

Ingredient	% (w/v) Composition
Bacto-tryptone	1.0
Yeast extract	0.1
Agar	1.0
Experimental conditions; prepared in filtered seawater; 120 °C/15 min.	

After Anderson *et al.* (1974).

Table 9.4 *Nutrient broth used for the mass cultivation of the antibiotic-producing marine purple-pigmented* Chromobacterium

Ingredient	% (w/v) Composition
Bacto-peptone	0.5
Yeast extract	0.1
Sodium bromide	0.05
Experimental conditions; prepared in seawater	

After Anderson *et al.* (1974).

Table 9.5 *Solid medium used for the mass cultivation of cells of* Chromobacterium *sp.*

Ingredient	% (w/v) Composition
Bacto-tryptone	1.0
Yeast extract	0.1
Sodium bromide	0.05
Agar	Not mentioned
Experimental conditions; prepared in seawater	

After Anderson *et al.* (1974).

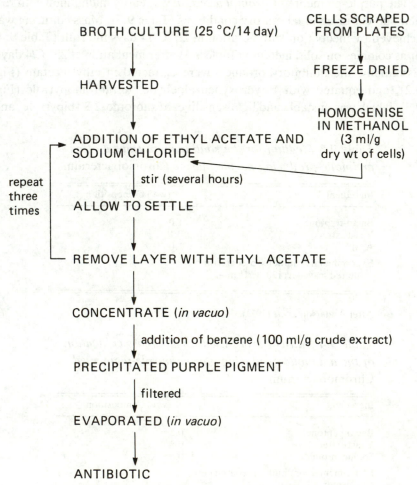

Fig. 9.2 Diagrammatic representation of the isolation of antibiotics from *Chromobacterium* sp. (after Anderson *et al.*, 1974).

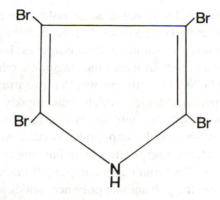

Fig. 9.3 Tetrabromopyrrole.

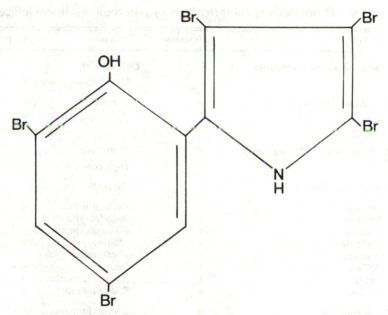

Fig. 9.4 2-(2'-hydroxy-3',5'-dibromophenyl)-3,4,5-tribromopyrrole.

2-(2'-hydroxy-3',5'-dibromophenyl)-3,4,5-tribromopyrrole (Fig. 9.4). The latter inhibited *Staphylococcus aureus*, but the greatest antibiotic activity was demonstrated by tetrabromopyrrole which antagonised the yeast *Candida albicans*, as well as *Escherichia coli*, *Pseudomonas aeruginosa* and *Staphylococcus aureus*.

The significance of brominated antibiotics increased as a result of work by a French group. Thus using culture collection strains of marine purple-pigmented *Alteromonas luteoviolaceus* (Table 9.6), Gauthier and Flateau

(1976) succeeded in isolating polyanionic polysaccharides as well as two brominated anti-bacterial compounds. Marine 2216E agar overlayered with cellophane sheets was used for cultivation purposes. Inhibitory compounds were extracted by a variety of techniques including ethanol with fractionation on Sephadex G-200. The outcome was that two strains produced inhibitory polyanionic polysaccharides, which were weakly bound to the cells, and diffused in the growth medium, and two intracellular brominated (with 2.9% bromine) anti-bacterial compounds. Further work established that these compounds affected respiration by enhancing oxygen uptake and by producing peroxides. The brominated compounds inhibited *Staphylococcus epidermidis* whereas the polyanionic polysaccharides antagonised *Bacillus*

Table 9.6 *Distinguishing characteristics of* Alteromonas luteoviolaceus

Character	Response	Character	Response
Presence of Gram-negative rods	+	*Utilisation of:*	
		N-acetylglucosamine	+
Production of purple pigment	+	*L*-α-alanine	+
		Calcium lactate	+ (some strains)
Motile by a single polar flagellum	+	*D*-fructose	−
		D-glucose	+
Oxidative metabolism	+	Glycerol	−
Production of:		Erythritol	−
Amylase	+	*m*-hydroxybenzoate	−
Catalase	+	α-ketoglutamate	−
Chitinase	−	Maltose	+
Gelatinase	+	*D*-mannose	−
DN'ase	+	Lactose	−
Lipase	+	Sodium succinate	−
Oxidase	+	Sodium citrate	−
Nitrate reduction	−	Sodium fumarate	−
Growth at:		*D*-sorbitol	−
4 °C	Weak	Sucrose	−
35 °C	+	*DL*-malate	−
40 °C	−	*L*-threonine	+
Requirement for organic growth factors	+	Trehalose	+
		L-tyrosine	+ (some strains)
Production of polyanionic antibiotics	+	*G+C content of DNA of 40–43 moles %*	+

After Baumann *et al.* (1984).

firmus, Staphylococcus epidermidis and *Staphylococcus aureus* with lesser activity against *Morganella morganii* and *Shigella dysenteriae*.

Seemingly, brominated compounds are common among marine Gram-negative bacteria insofar as recovery is possible by use of simple non-selective techniques. However, the relationship of *Chromobacterium* sp., *Alteromonas luteoviolaceus* and *P. bromoutilis* should be resolved in order to determine whether or not the antibiotics are produced by one or more bacterial genera. Thence, selective enrichment techniques could be developed to recover many more strains for antibiotic screening programmes with a view to discovering different inhibitory compounds. This could be extremely profitable insofar as the majority of screening programmes to date highlight Gram-positive bacteria, notably actinomycetes. Yet in the marine environment, the Gram-negative component of the microflora may possess commercially exploitable antibiotics. In addition to the organisms already mentioned, a marine vibrio, termed *Alginovibrio aquatilis* (this is not included in the *Approved Lists of Bacterial Names*, or the supplements), has been reported to inhibit mucoid strains of *Pseudomonas aeruginosa* by means of a heat stable isoamyl-alcohol soluble compound (Doggett, 1968).

In an interesting study, Wratten *et al.* (1977) recovered antibiotic-producing Gram-negative, yellow-pigmented, motile rods, which produced catalase, indole and oxidase, required salt, and gave negative responses to the methyl red test and Voges Proskauer reaction. The organisms were considered to represent marine pseudomonads, although a relationship to *Alteromonas* should not be ruled out. Although the precise method of isolation was not published, culturing was possible on trays of nutrient agar (comprises 0.5% (w/v) peptone, 0.1% (w/v) yeast extract, 1.0% (w/v) agar, prepared in filtered seawater). After incubation at room temperature for four days, antibiotics were extracted in ethyl acetate, with drying over anhydrous sodium sulphate, followed by chromatographic separation. The pseudomonad, which was inhibitory to *Candida albicans, Staphylococcus aureus, Vibrio anguillarum* and *V. harveyi*, produced two novel antibiotics, namely 2-*n*-pentyl-4-quinolinol (Fig. 9.5) and 2-*n*-heptyl-4-quinolinol (Fig. 9.6). This class of antimicrobial compound has great value in human and veterinary medicine, insofar as plasmid-mediated resistance does not generally occur. Therefore it may be anticipated that examination of the antimicrobial activity of quinolinols will feature firmly in future research programmes.

Interest in the production of antibiotics by marine Gram-negative bacteria has continued. Lemos *et al.* (1985) examined the microflora from seaweeds (*Enteromorpha intestinalis, E. compressa, Fucus ceranoides, Pelvetia canaliculata* and *Ulva lactuca*), which were collected in Spain from the

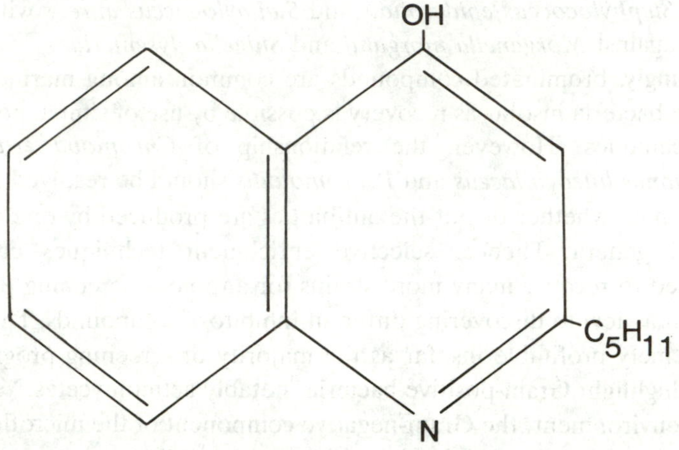

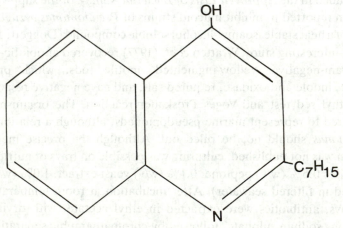

Fig. 9.5 2-*n*-pentyl-4-quinolinol.

Fig. 9.6 2-*n*-heptyl-4-quinolinol.

intertidal zone at low tide. The seaweed was rinsed in sterile seawater to remove superfluous organisms, and a one cm^2 area was swabbed. This material was spread over the surface of marine 2216E agar plates, with incubation at 20 °C for 7 days. Antibacterial activity was demonstrated in 38/224 (17%) isolates derived from 62 samples of seaweed, with the majority originating from *E. intestinalis*. Characteristically, the isolates were Gram-negative, motile, pigmented (green, orange or purple) rods, which produced catalase and oxidase. They were considered to be pseudomonads, possibly related to *Alteromonas*. The inhibitory agent was described as a low molecular weight (<2000 daltons) compound. In addition, Imada *et al*. (1986)

recovered low molecular weight (1500–1700 daltons) peptides, termed marinostatins, from *Alteromonas*.

It should not be misconstrued that the majority of marine antibiotic-producers comprise Gram-negative rods, insofar as several interesting compounds have stemmed from work with Gram-positive bacteria. Indeed, Baarn *et al.* (1966) indicated that the majority of marine antibiotic-producers comprised Gram-positive bacteria. Thus using marine 2216E agar and Czepaks medium supplemented with 3% (w/v) sodium chloride (for marine actinomycetes), these workers recovered 60 antibiotic-producing cultures from seawater. Of these organisms, 45 (75%) were endospore-forming Gram-positive rods (presumptive *Bacillus* spp.), 11 (19%) were Gram-positive cocci, two (3%) were *Streptomyces* spp. and the remaining two (3%) cultures comprised Gram-negative rods. Certainly, this antibiotic-producing ability of Gram-positive bacteria has been admirably developed by Okami (1986).

From an isolate (SS–20) of *Streptomyces griseus* recovered from shallow sea mud in Japan, a novel antibiotic was discovered, for which the name of *Aplasmomycin* was coined. Antibiotic production was maximal in broth supplemented with 1% (w/v) glucose, 1.5% (w/v) sodium chloride and powdered *Laminaria*, termed Kobucha. From this, the antibiotic was extracted (Fig. 9.7) as colourless needles with a melting temperature of 283–285 °C. It was soluble in organic solvents, but only slightly in water. The molecular formula was deduced to be $C_{40}H_{60}O_{14}BNa$ (Fig. 9.8). Two minor products were also produced, namely aplasmomycin *B* with an acetyl group at either the C-9 or C-9′ position and aplasmomycin *C* with acetyl groups at both of the aforementioned carbon positions. Subsequent investigation revealed that the antibacterial activity of aplasmomycins *B* and *C* was nearly equal to and weaker than that of aplasmomycin, respectively. This activity was pronounced against mycobacteria and other Gram-positive bacteria (inhibitory dose = 0.78–6.25 μg/ml) *in vitro*, and against *Plasmodium berghei* (the causal agent of malaria) in mice. The drug was duly named from this latter activity.

Okami (1986) described a novel aminoglycoside antibiotic, *Istamycin*, from an actinomycete which was recovered from shallow sea mud in Sagami Bay, Japan. Isolation was achieved by suspending mud samples in seawater (1:4), and spreading 0.05 ml aliquots on the surface of maltose (1.0% w/v) yeast extract (0.4% w/v) agar, prepared in seawater and supplemented with antifungal (nystatin) and antibacterial (kanamycin) agents. After incubation at 27 °C for 4–10 days, a colony type was selected which produced pale blue aerial mycelia. Only glucose and inositol were utilised as the sole sources of carbohydrate on Pridham–Gottliebs agar. From this scant descrip-

BROTH CULTURE

↓

ADJUST TO pH 5.0 WITH HCl

| filtered to remove mycelia

EXTRACT FILTRATE WITH BUTYL ACETATE

| concentrate solvent layer (*in vacuo* at 40 °C)

↓

YELLOW VISCOUS LIQUID

↓

CHROMATOGRAPHY (neutral alumina)

| 9:1 mixture of benzene and ethyl acetate

↓

COLLECT PURIFIED FRACTIONS WITH ANTIMICROBIAL ACTIVITY

| concentrate (*in vacuo* at 40 °C)

PALE YELLOW POWDER

↓

CHROMATOGRAPHY (silicic acid)

| 9:1 mixture of n-hexane and ethyl acetate

COLLECT FRACTIONS WITH ANTIMICROBIAL ACTIVITY

| shake with 2N NaOH

↓

WASH WITH WATER

| dehydrate with anhydrous sodium sulphate

↓

EVAPORATE TO REMOVE SOLVENT (*in vacuo* at 40 °C)

↓

CRYSTALLISE ANTIBIOTIC FROM METHANOL – WATER MIXTURE

Fig. 9.7 Steps in the recovery of aplasmomycins from marine isolates of *Streptomyces griseus* (after Okami, 1986).

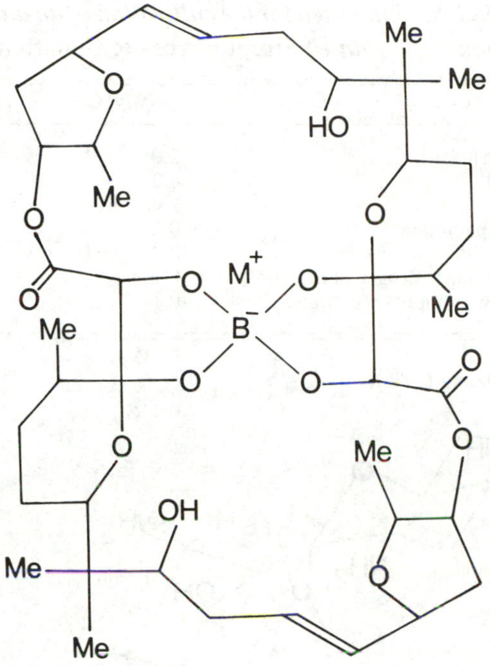

Fig. 9.8 Aplasmomycin.

tion, the organism was equated with *Streptomyces*, and elevated into a new species as *Streptomyces tenjimariensis*, although the name is not included in the *Approved Lists of Bacterial Names* (Skerman *et al.*, 1980) or their supplements, and therefore lacks taxonomic validity. An isolate (SS-939), which was recovered in 1978, was cultured aerobically (Table 9.7) at 27 °C for three days with agitation (300 rpm). The culture was filtered at pH 2.0, and the filtrate was neutralised prior to adsorption of the antibiotics on to a column of Amberlite 1RC-50. Following elution with 1 N aqueous ammonia, the extracts were concentrated, and fractionated by column chromatography. The antibiotics consisted of an amino sugar and an aminocyclitol moiety, and were considered to belong to the fortimicin–sporaricin group, being named as istamycin *A* and *B* (Figs. 9.9, 9.10). These aminoglycoside antibiotics inhibited a wide range of Gram-positive and Gram-negative bacteria, including many of the organisms resistant to other aminoglycosides. Generally, the inhibitory activity of istamycin *B* was two to four-fold greater than that of istamycin *A*. Clearly, it is worthwhile searching for other novel aminoglycoside antibiotics among streptomycetes of marine origin.

Table 9.7 *Medium used for the cultivation of an antibiotic-producing strain of* Streptomyces tenjimariensis

Ingredient	% (w/v) Composition
Soybean meal	2.0
Glucose	0.2
Starch	2.0
Sodium palmitate	0.2
Sodium chloride	0.3
Dipotassium hydrogen phosphate	0.1
Magnesium sulphate (hydrated)	0.1

After Okami (1986).

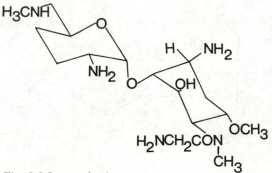

Fig. 9.9 Istamycin *A*.

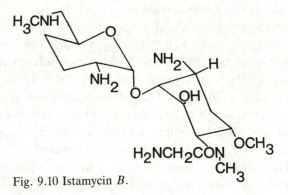

Fig. 9.10 Istamycin *B*.

9.1.2   *Antiviral compounds*

Although the majority of studies have highlighted antibacterial agents, there are a small number of articles dealing with the production of antiviral compounds by marine micro-organisms. In this respect, Magnussen *et al.* (1967) detailed the effect of marine bacteria, notably *Vibrio marinus*, on inactivation of enteroviruses, and Katzenelson (1978) reported the antiviral activity of marine *Flavobacterium*. Thus it seems hardly surprising that Toranzo *et al.* (1982) attributed antipoliovirus activity to marine *Pseudomonas* or *Vibrio*. This activity is undoubtedly contained in culture supernatants but it remains for further work to elucidate the exact chemical structure of the compound(s). Again, this is an area worthy of further study.

9.1.3 *Antitumour compounds*

Okami (1986) discussed the isolation from a marine flavobacterium of a polysaccharide with marked activity in mice against Sarcoma-180 solid tumour virus (S-180). From seaweed, which was collected in Sagami Bay, Japan, bacterial isolations were carried out on a non-selective, nutrient medium (Table 9.8) with incubation at 27 °C for 1–3 days. Isolated colonies were examined for the ability to produce polysaccharide on a medium containing sugar (Table 9.9) after incubation at 27 °C for 2 days. Any polysaccharide was extracted chemically (Fig. 9.11). In experiments with mice, the crude polysaccharide protected against the ravages of S-180 virus. Of 500 bacterial isolates, 20% produced polysaccharide but only 5% pos-

Table 9.8 *Medium used for the recovery of marine organisms which produced antitumour polysaccharide*

Ingredient	% (w/v) Composition
Polypeptone	0.2 –0.5
Yeast extract	0.03–0.1
Agar	1.8
Experimental conditions; prepared in artificial seawater; pH 7.2	

After Okami (1986).

Table 9.9 *Medium used for the production of antitumour polysaccharide by marine bacteria*

Ingredient	% (w/v) Composition
Polypeptone	0.5
Yeast extract	0.1
Glucose or sucrose	0.6
Experimental conditions; prepared in artificial seawater; pH 7.2	

After Okami (1986).

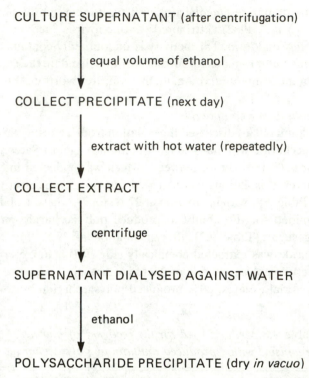

CULTURE SUPERNATANT (after centrifugation)

equal volume of ethanol

COLLECT PRECIPITATE (next day)

extract with hot water (repeatedly)

COLLECT EXTRACT

centrifuge

SUPERNATANT DIALYSED AGAINST WATER

ethanol

POLYSACCHARIDE PRECIPITATE (dry *in vacuo*)

Fig. 9.11 Steps in the recovery of antitumour polysaccharide from marine bacteria (after Okami, 1986).

sessed inhibitory activity in mice. A seawater-requiring strain, MP-55, produced a water-soluble compound, named *Marinactan* (contains 69.7% glucose, 19.8% mannose, 10.8% fucose), which produced complete regression of tumours. Purified marinactan gave a 75–95% inhibition in the development of tumours in some animals. The producer organism was identified as *Flavobacterium uliginosum*, although this taxon appears to comprise marine cytophagas, related to *Cytophaga lytica* (Bauwens and De Ley, 1981). Clearly, other marine organisms must produce antitumour compounds so a detailed search is justified.

9.2   Enzymes

The commercial value of enzymes has increased substantially, with uses including the confectionery (lactases) and detergent (proteases) industries. The search for novel or improved enzymes is a lucrative business for which the marine environment should not be overlooked. Already, it has been established that some marine bacteria produce copious quantities of alginate lyases and chitinases, which may warrant commercial exploitation.

For that matter, deep-sea thermophilic bacteria may provide useful sources of heat-stable enzymes (Deming, 1986). Certainly, the ability of marine micro-organisms to degrade complex polymers has been recognised. Moreover, the usefulness of glucan-degrading bacteria has been explored. This has significance in dental caries, insofar as glucan is produced by *Streptococcus mutans*, which is one of the possible causal agents of caries. Here, insoluble glucan, produced by oral streptococci, sticks to the surface of the teeth, and hence leads to dental plaque. Okami (1986) recovered a useful isolate of *Bacillus circulans* (No. MT-G2), from marine mud in Tokyo Bay, which produced an enzyme capable of hydrolysing glucan. Enzyme production, which was induced by the presence of glucan in the medium, was maximal after six days. Extraction was achieved with water, and thence by salting with ammonium sulphate. Fractionation was by chromatography on DEAE-Sephadex eluted with potassium chloride, followed by further chromatography on DE-32 cellulose which was eluted with acetate buffer. Finally, gel filtration was with Sephadex G-150. From this procedure, a novel enzyme was recovered, which degraded glucans consisting of α-1,3 and/or α-1,6 linkages. The optimum temperature and pH for activity were 35 °C and 6.2–6.7, respectively. It is speculative whether or not this enzyme could be used for the prevention/reduction of dental caries but its incorporation into toothpastes for this purpose is a possibility. With such success, the search for more enzymes seems likely to proceed.

9.3   Surfactants

An isolate of *Acinetobacter* (No. RAG-1), recovered from seawater, produced a surfactant, which has been exploited commercially, as the product Emulsan (Patel and Hou, 1983). The surfactant polymer comprised a polysaccharide moiety of *N*-acetyl-*D*-galactosamine, *N*-acetylaminouronic acid and an amino sugar. α- and β-hydroxydodecanoic acids were found to be esterified to the sugars (Zosin *et al.*, 1982). The product has been marketed with great success as a cleansing agent for oil tankers and oil storage tanks. Another trait of the compound concerns its ability to complex with uranium. Therefore, there is the prospect that Emulsan may be useful for the recovery of uranium from waste material (Zosin *et al.*, 1983). With such success, there is every reason to support detailed searches for yet more useful microbial products.

9.4   Other potentially useful microbial products

It does not require profound judgement to deduce areas where marine micro-organisms could also contribute to biotechnology. The production by marine organisms of useful polymers, such as polyhydroxy-acid which is of commercial importance (Staley and Stanley, 1986), and vitamins is well established. Bacterial polymers have also been implicated in inducing settling of marine invertebrate larvae (Bonar *et al.*, 1986). Neumann (1979) attributed the property in *Vibrio* to ultrafiltrates of 1000–10 000 daltons molecular weight. Kirchman *et al.* (1982a,b) determined that in *Pseudomonas*, a polysaccharide or glycoprotein was responsible for inducing metamorphosis in the marine polychaete, *Janua brasiliensis*. It is tempting to speculate that the production of such compounds could have value in aquaculture or re-stocking programmes. There is also the potential for production of single cell protein. In this context, the marine yeast *Candida maris* grows well on methanol (Phaff, 1986). So, from a cheap substrate, the production of protein is realistic. Alternatively, there is the possibility of energy production, such as hydrogen, from microbial biomass, again using marine microbes. Finally, marine bacteria may harbour plasmids, which could be of value to molecular biologists. Thus, it may be deduced that marine microbiology offers considerable promise for biotechnology. The impetus has started, so let us see how long it takes for the efforts to bear greater fruit.

10   Future developments

Predicting the future is a notoriously uncertain affair, with developments in one area greatly influencing other apparently distant disciplines. Who would have been able to foresee the tremendous advances in molecular genetics, immunology (i.e. establishment of monoclonal antibody technology) and biotechnology, and their pronounced effect on other facets of biology. Taking marine microbiology, it may appear to some that there has been a dearth of significant advances in knowledge since the pioneering work of Zobell and co-workers. Instead, it may be argued that there has been much elaborate refining of long established ideas. Interestingly, a critical appraisal of the literature for the four decades from 1946 – the year of publication of Zobell's treatise on marine microbiology – to 1986, reveals a progression of fashionable trends.

Great excitement was expressed in some quarters over the work on food webs. To the layman, the potential uptake and transmission of radio-isotopes from unicellular to multicellular organisms assumes significance in view of the accidental release of radioactive compounds from nuclear power stations such as Chernobyl, and the associated re-processing plants, e.g. Sellafield (formerly known as Windscale).

The true relevance of the concept of dormancy has still to be realised, particularly with regard to the survival of pathogens in the marine environment. Nevertheless although there is continued research on somewhat hackneyed ideas, new areas are slowly emerging.

The exploration for oil and the unfortunate but resultant pollution by hydrocarbons prompted some excellent work during the 1970s and early 1980s. However, during the next millenium, the oil industry as it exists at present seems doomed to falter. Yet, the problems of pollution will not diminish, especially if human populations continue to increase. Instead, the troubles of release of human pathogens and radioactive compounds into the aquatic environment may assume even greater importance. Consequently, it is predicted that pollution research will gain in momentum. In short, applied aspects of science seem destined for greater support over the much neglected basic studies. This may be further illustrated by the current developments in biotechnology, and the continued search for novel pharmaceutical compounds, enzymes and substrates. The signs are that the marine environment is slowly being appreciated as a potential source of such compounds. Therefore, it is anticipated that there will be an injection of much needed finance into the discipline of marine biotechnology in order to develop the sorely needed selective isolation methods for commercially useful organisms and screening programmes for potentially interesting compounds. Perhaps, this may also encourage taxonomy and ecology, so long neglected for sufficient resources.

Finally with the continued expansion of aquaculture, particularly mariculture, the topic of fish and shellfish disease research must surely increase. Nevertheless only time will reveal where the actual future developments will be made.

References

Chapter 1

Grant Gross, M. (1976). *Oceanography* 3rd edn. Charles E. Merrill: Columbus, Ohio.

Holland, H.D. (1978). *The Chemistry of the Atmosphere and Oceans.* Wiley: New York.

Naumann, E. (1917). Beitrag zur Kenntnis des Teichnannoplankton, 11. Uber das Neuston der Süsserwassers. *Biologische Zentrablatt*, **37**, 98–106.

Pytkowicz, R.M. and Kester, D.R. (1971). The physical chemistry of seawater. In *Annual Review of Oceanographic Marine Biology* H. Barnes (ed), pp. 11–60. Harper & Row: New York.

Sieburth, J.McN. (1979). *Sea Microbes.* Oxford University Press: New York.

Chapter 2

Anon. (1970). *Recommended Procedures for the Examination of Sea Water and Shellfish, 4th edn.* American Public Health Association: Washington, DC.

Anon. (1975). *Standard Methods for the Examination of Water and Wastewater, 14th edn.* American Public Health Association: Washington, DC.

Austin, B. (1983). Bacterial microflora associated with a coastal, marine fish–rearing unit. *Journal of the Marine Biological Association, UK*, **63**, 585–92.

Austin, B. (1987). Deep sea microbiology. In *Methods in Aquatic Bacteriology.* B. Austin (ed.), Wiley: Chichester (in press).

Austin, B., Allen, D.A., Zachary, A., Belas, M.R. and Colwell, R.R. (1979). Ecology and taxonomy of bacteria attaching to wood surfaces in a tropical harbor. *Canadian Journal of Microbiology*, **25**, 446–61.

Biebl, H. and Pfennig, N. (1978). Growth yields of green sulfur bacteria in mixed cultures with sulfur and sulfate reducing bacteria. *Archives of Microbiology*, **117**, 9–16.

Conrad, R. and Schütz, H. (1987). Methods to study methanogenic bacteria and methanogenic activities in aquatic environments. In *Methods in Aquatic Bacteriology.* B. Austin (ed), Wiley: Chichester (in press).

Hamilton, R.D. and Holm-Hansen, O. (1967). Adenosine triphosphate content of marine bacteria. *Limnology and Oceanography*, **12**, 319–24.

Holm-Hansen, O. and Booth, C.R. (1966). The measurement of adenosine triphosphate in the ocean and its ecological significance. *Limnology and Oceanography*, **11**, 510–19.

Hungate, R.E. (1969). A roll tube method for the cultivation of strict anaerobes. In *Methods in Microbiology*, J.R. Norris and D.W. Ribbons (eds), pp. 117–32. Academic Press: London.

Imhoff, J.F. (1987). Anoxygenic phototrophic bacteria. In *Methods in Aquatic Bacteriology*. B. Austin (ed), Wiley: Chichester (in press).

Karl, D.M. (1980). Cellular nucleotide measurements and applications in microbial ecology. *Microbiological Reviews*, **44**, 739–96.

Kester, D.R., Duedall, I.W., Connors, D.N. and Pytkowicz, R.M. (1967). Preparation of artificial seawater. *Limnology and Oceanography*, **12**, 176–9.

Kobayashi, T., Enomoto, S., Sakazaki, R. and Kuwahara, S. (1963). A new selective medium for pathogenic vibrios, TCBS agar (modified Nakanishi's agar). *Japanese Journal of Bacteriology*, **18**, 387–91.

Olson, B.H. (1978). Enhanced accuracy of coliform testing in seawater by a modification of the Most Probable Number method. *Applied and Environmental Microbiology*, **36**, 438–44.

Parsons, T.R., Maita, Y. and Lalli, C.M. (1984). *A Manual of Chemical and Biological Methods for Seawater Analysis*. Pergamon Press: Oxford.

Pfennig, N. and Trüper, H.G. (1981). Chapter 16. Isolation of members of the families Chromatiaceae and Chlorobiaceae. In *The Prokaryotes A Handbook on Habitats, Isolation, and Identification of Bacteria*, M.P. Starr, H. Stolp, H.G. Trüper, A. Balows and H.G. Schlegel (eds), vol.1., pp. 279–89. Springer-Verlag: Berlin.

Rodina, A.G. (1972). *Methods in Aquatic Microbiology* (translated by Colwell, R.R. and Zambruski, M.S.). University Park Press: Baltimore.

Schneider, J. and Rheinheimer, G. (1987). Isolation methods. In *Methods in Aquatic Bacteriology*. B. Austin (ed), Wiley: Chichester (in press).

Seliger, H.H. and McElroy, W.D. (1960). Spectral emission and quantum yield of firefly bioluminescence. *Archives of Biochemistry and Biophysics*, **88**, 136–141.

Sieburth, J.McN. (1979). *Sea Microbes*. Oxford University Press: New York.

Simidu, U. (1974). Improvement of media for enumeration and isolation of heterotrophic bacteria in seawater. In Effect of the Ocean Environment on Microbial Activities, R.R. Colwell and R.Y. Morita (eds), pp. 249–57. University Park Press: Baltimore.

Stal, L.J., van Gemerden, H. and Krumbein, W.E. (1984). The simultaneous assay of chlorophyll and bacteriochlorophyll in natural microbial communities. *Journal of Microbiological Methods*, **2**, 295–306.

Sullivan, J.D. and Watson, S.W. (1974). Factors affecting the sensitivity of *Limulus* lysate. *Applied Microbiology*, **28**, 1023–6.

Trüper, H.G. (1970). Culture and isolation of phototrophic sulphur bacteria from the marine environment. *Helgoländer wissenschaftliche Meeresuntersuchungen*, **20**, 6–16.

Watson, S.W., Novitsky, T.J., Quinby, H.L. and Valois, F.W. (1977). Determination of bacterial numbers and biomass in the marine environment. *Applied and Environmental Microbiology*, **33**, 940–54.

Watson, S.W. and Hobbie, J.E. (1979). Measurement of bacterial biomass as lipopolysaccharide. In *Native Aquatic Bacteria: Enumeration, Activity and Ecology*, J.W. Costerton, and R. R. Colwell (eds), pp. 82–8. American Society for Testing and Materials, ASTM STP 695: Philadelphia.

White, D.C. (1983). Analysis of micro-organisms in terms of quantity and activity in natural environments. In *Microbes in their Natural Environments*, J.H. Slater, R.Whittenbury and J.W.T.Wimpenny (eds), pp. 37–66. Cambridge University Press: Cambridge.

White, D.C., Bobbie, R.J., King, J.D., Nickels, J. and Amoe, P. (1979). Lipid analysis of sediments for microbial biomass and community structure. In *Methodology for Biomass Determinations and Microbial Activities in Sediments*, C.D. Litchfield and P.L. Seyfried (eds), pp. 87–103. American Society for Testing and Materials ASTM STP 673: Philadelphia.

Zobell, C.E. (1941). Studies on marine bacteria. 1. The cultural requirements of heterotrophic aerobes. *Journal of Marine Research*, **4**, 42–75.

Chapter 3

Ahearn, D.G. and Crow, S.A. (1986). Fungi and hydrocarbons in the marine environment. In *The Biology of Marine Fungi*, S.T. Moss (ed), pp. 11–18. Cambridge University Press: Cambridge.

Alldredge, A.L., Cole, J.J. and Caron, D.A. (1986). Production of heterotrophic bacteria inhabiting macroscopic organic aggregates (marine snow) from surface waters. *Limnology and Oceanography*, **31**, 66–78.

Allen, W.E. (1941). Depth relationships of plankton diatoms in sea water. *Journal of Marine Research*, **4**, 107–11.

Andersen, P. and Sørensen, H.M. (1986). Population dynamics and trophic coupling in pelagic micro-organisms in eutrophic coastal waters. *Marine Ecology–Progress Series*, **33**, 99–109.

Andersson, A., Larsson, U. and Hagström, Å. (1986). Size-selective grazing by a microflagellate on pelagic bacteria. *Marine Ecology–Progress Series*, **33**, 51–7.

Araki, T. and Kitamikado, M. (1978). Distribution of mannan-degrading bacteria in aquatic environments. *Bulletin of the Japanese Society of Scientific Fisheries*, **44**, 1135–9.

Ashby, R.E. and Rhodes-Roberts, M.E. (1976). The use of analysis of variance to examine the variations between samples of marine bacterial populations. *Journal of Applied Bacteriology*, **41**, 439–51.

Austin, B. (1983). Bacterial microflora associated with a coastal, marine fish-rearing unit. *Journal of the Marine Biological Association UK*, **63**, 585–92.

Bonde, G.J. (1975). The genus *Bacillus*. *Danish Medical Bulletin*, **22**, 41–61.

Bonde, G.J. (1976). The marine *Bacillus*. *Journal of Applied Bacteriology*, **41**, vii.

Bunch, J.N. and Harland, R.C. (1976). *Biodegradation of crude petroleum by the indigenous microbial flora of the Beaufort Sea*. Beaufort Sea Project Technical Report, No. 10. Information Canada, Victoria, BC.

Campbell, L. and Carpenter, E.J. (1986). Estimating the grazing pressure of heterotrophic nanoplankton on *Synechococcus* spp. using the sea water dilution and selective inhibitor techniques. *Marine Ecology–Progress Series*, **33**, 121–9.

Carlucci, A.F. (1974). Nutrients and microbial response to nutrients in seawater. In *Effect of the Ocean Environment on Microbial Activities*, R.R. Colwell and R.Y. Morita (eds), pp. 245–8. University Park Press: Baltimore.

Caron, D.A. (1983). Technique for enumeration of heterotrophic and phototrophic nanoplankton, using epifluorescence microscopy, and comparison with other techniques. *Applied and Environmental Microbiology*, **46**, 491–8.

Caron, D.A., Davis, P.G., Madin, L.P. and Sieburth, J.McN. (1982). Heterotrophic bacteria and bacteriovorous protozoa in oceanic micro-aggregates. *Science (Washington, DC)*, **218**, 795–7.

Chan, K.Y., Woo, M.L., Lo, K.W. and French, G.L. (1986). Occurrence and distribution of halophilic vibrios in subtropical coastal waters of Hong Kong. *Applied and Environmental Microbiology*, **52**, 1407–11.

Crow, S.A., Ahearn, D.G. and Cook, W.L. (1975). Densities of bacteria and fungi as determined by a membrane–absorption procedure. *Limnology and Oceanography*, **20**, 644–6.

Dahle, A.B. and Laake, M. (1982). Diversity dynamics of marine bacteria studied by immunofluorescent staining on membrane filters. *Applied and Environmental Microbiology*, **43**, 169–76.

Daley, R.J. and Hobbie, J.E. (1975). Direct counts of aquatic bacteria by a modified epifluorescence technique. *Limnology and Oceanography*, **20**, 875–82.

Davis, P.G. and Sieburth, J.McN. (1984). Estuarine and oceanic microflagellate predation of actively growing bacteria: estimation by frequency of dividing–divided bacteria. *Marine Ecology–Progress Series*, **19**, 237–46.

El Hag, A.G.D. and Fogg, G.E. (1986). The distribution of coccoid blue-green algae (cyanobacteria) in the Menai Straits and the Irish Sea. *British Phycology Journal*, **21**, 45–54.

Erkenbrecher, C.W. and Stevenson, L.H. (1975). The influence of tidal flux on microbial biomass in salt marsh creeks. *Limnology and Oceanography*, **20**, 618–25.

Ezura, Y., Daiku, K., Tajima, K., Kimura, T. and Sakai, M. (1974). Seasonal difference in bacterial counts and heterotrophic bacterial flora in Akkeshi Bay. In *Effect of the Ocean Environment on Microbial Activities*, R.R. Colwell and R.Y. Morita (eds), pp. 112–23. University Park Press: Baltimore.

Fell, J.W. (1974). Distribution of yeasts in the water masses of the southern oceans. In *Effect of the Ocean Environment on Microbial Activities*, R.R. Colwell and R.Y. Morita (eds), pp. 510–23. University Park Press: Baltimore.

Ferguson, R.L. and Rublee, R. (1976). Contribution of bacteria to standing crop of coastal plankton. *Limnology and Oceanography*, **22**, 141–5.

Fogg, G.E. (1966). The extracellular products of algae. *Oceanography and Marine Biology Annual Review*, **4**, 195–212.

Fogg, G.E. (1987). Marine planktonic cyanobacteria. In *Cyanobacteria: Current Research*. C. Van Baalen and P. Fay (eds), (in press).

Fuhrman, J.A. and McManus, G.B. (1984). Do bacteria-sized marine eukaryotes consume significant bacterial production? *Science (Washington, DC)*, **224**, 1257–60.

Grüttner, H. and Jensen, K. (1983). Effects of chronic oil pollution from refinery effluent on sediment microflora in a Danish coastal area. *Marine Pollution Bulletin*, **14**, 436–59.

Gunn, B.A., Singleton, F.L., Peele, E.R. and Colwell, R.R. (1982). A note on the isolation and enumeration of Gram positive cocci from marine and estuarine waters. *Journal of Applied Bacteriology*, **53**, 127–9.

Hagler, A.N. and Ahearn, D.G. (1986). Ecology of aquatic yeasts. In *The Yeasts*. A.H. Rose (ed), Academic Press: London.

Hanson, R.B., Shafer, D., Ryan, T., Pope, D.H. and Lowery, H.K. (1983). Bacterioplankton in Antarctic Ocean waters during late austral winter: abundance, frequency of dividing cells, and estimates of production. *Applied and Environmental Microbiology*, **45**, 1622–32.

Hasle, G.R. (1950). Phototactic vertical migrations in marine dinoflagellates. *Oikos*, **2**, 162–75.

Hines, M.E. and Buck, J.D. (1982). Distribution of methanogenic and sulfate-reducing bacteria in near–shore marine sediments. *Applied and Environmental Microbiology*, **43**, 447–53.

Honjo, S. and Okado, H. (1974). Community structure of coccolithophores in the photic layer of the mid-Pacific. *Micropaleontology*, **20**, 209–30.

Ishida, Y., Eguchi, M. and Kadota, H. (1986). Existence of obligately oligotrophic bacteria as a dominant population in the South China Sea and the West Pacific Ocean. *Marine Ecology–Progress Series*, **30**, 197–203.

Jannasch, H.W. and Jones, G.E. (1959). Bacterial populations in seawater as determined by different methods of enumeration. *Limnology and Oceanography*, **4**, 128–39.

Kadota, H. (1959). Cellulose decomposing bacteria in the sea. In *Marine Boring and Fouling Organisms*, D.L. Ray (ed), pp. 332–40. University of Washington Press, Seattle.

Kaneko, T., Roubal, G. and Atlas, R.M. (1978). Bacterial populations in the Beaufort Sea. *Arctic*, **31**, 91–107.

Kaplan, I.R. and Friedmann, A. (1970). Biological productivity in the Dead Sea. Part 1. Microorganisms in the water column. *Israel Journal of Chemistry*, **8**, 513–28.

Karl, D.M. (1978). Distribution, abundance, and metabolic states of microorganisms in the water column and sediments of the Black Sea. *Limnology and Oceanography*, **23**, 936–49.

Khiyama, H.M. and Makemøson, J.C. (1973). Sand beach bacteria: enumeration and characterization. *Applied Microbiology*, **26**, 293–7.

Kjelleberg, S. and Håkansson, N. (1977). Distribution of lipolytic, proteolytic, and amylolytic marine bacteria between the lipid film and the subsurface water. *Marine Biology*, **39**, 103–9.

Kogure, K., Simidu, U. and Taga, N. (1979). A tentative direct microscopic method for counting living marine bacteria. *Canadian Journal of Microbiology*, **25**, 415–20.

Kogure, K., Simidu, U. and Taga, N. (1980a). Effect of phyto- and zooplankton on the growth of marine bacteria in filtered seawater. *Bulletin of the Japanese Society of Scientific Fisheries*, **46**, 323–6.

Kogure, K., Simidu, U. and Taga, N. (1980b). Distribution of viable marine bacteria in neritic seawater around Japan. *Canadian Journal of Microbiology*, **26**, 318–23.

Kohlmeyer, J. (1983). Geography of marine fungi. *Australian Journal of Botany, Supplement Series*, **10**, 67–76.

Kohlmeyer, J. and Kohlmeyer, E. (1979). *Marine Mycology: The Higher Fungi*. Academic Press: New York.

Kriss, A.E. (1963). *Marine Microbiology (Deep Sea)*. (Translated by J.M. Shewan and Z. Kabata), Oliver and Boyd: Edinburgh.

Krumbein, W.E. (1971). Sediment microbiology and grain–size distribution as related to tidal movement, during the first mission of the West German undersea laboratory 'Helgoland'. *Marine Biology*, **10**, 101–12.

Lackey, J.B. and Clendinning, K.A. (1965). Ecology of the microbiota of San Diego Bay, California. *Transactions of the San Diego Society of Natural History*, **14**, 9–40.

Li, W.K.W., Subba Rao, D.V., Harrison, W.G., Smith, J.C., Cullen, J.J., Irwin, B. and Platt, T. (1983). Autotrophic picoplankton in the tropical Ocean. *Science (Washington DC)*, **219**, 292–5.

Litchfield, C.D. and Floodgate, G.D. (1975). Biochemistry and microbiology of some Irish Sea sediments: II. Bacteriological Analyses. *Marine Biology*, **30**, 97–103.

Maeda, M. and Taga, N. (1973). Deoxyribonuclease activity in seawater and sediment. *Marine Biology*, **20**, 58–63.

Maeda, M. and Taga, N. (1979). Chromogenic assay method of lipopolysaccharide (LPS) for evaluating bacterial standing crop in seawater. *Journal of Applied Bacteriology*, **47**, 175–82.

Mallory, L.M., Austin, B. and Colwell, R.R. (1977). Numerical taxonomy and ecology of oligotrophic bacteria isolated from the estuarine environment. *Canadian Journal of Microbiology*, **23**, 733–50.

Martin, Y.P. and Bianchi, M.A. (1980). Structure, diversity, and catabolic potentialities of aerobic heterotrophic bacterial populations associated with continuous cultures of natural marine phytoplankton. *Microbial Ecology*, **5**, 265–79.

Meadows, P.S. and Anderson, J.G. (1966). Microorganisms attached to marine and freshwater sand grains. *Nature (London)*, **198**, 610–11.

McAlice, B.J. (1970). Observations on the small scale distribution of estuarine phytoplankton. *Marine Biology*, **7**, 100–11.

McCallum, M.F. (1970). Aerobic bacterial flora of the Bahama Bank. *Journal of Applied Bacteriology*, **33**, 533–42.

Meyer-Reil, L.-A., Dawson, R., Liebezeit, G. and Tiedge, H. (1978). Fluctuations and interactions of bacterial activity in sandy beach sediments and overlying waters. *Marine Biology*, **48**, 161–71.

Meyers, S.P., Ahearn, D.G., Gunkel, W. and Roth, F.J. (1967). Yeasts from the North Sea. *Marine Biology*, **1**, 118–23.

Montagna, P.A. (1984). *In situ* measurement of meiobenthic grazing rates on sediment bacteria and edaphic diatoms. *Marine Ecology–Progress Series*, **18**, 119–130.

Moriarty, D.J.W., Pollard, P.C. and Hunt, W.G. (1985). Temporal and spatial variation in bacterial production in the water column over a coral reef. *Marine Biology*, **85**, 285–92.

Murphy, L.S. and Haugen, E.M. (1985). The distribution and abundance of phototrophic ultraplankton in the North Atlantic. *Limnology and Oceanography*, **30**, 47–58.

Novitsky, J.A. and Karl, D.M. (1985). Influence of deep ocean sewage outfalls on the microbial activity of the surrounding sediment. *Applied and Environmental Microbiology*, **50**, 1464–73.

Okami, Y. and Okazaki, T. (1978). Actinomycetes in marine environments. In

Nocardia and Streptomyces. Proceedings of the International Symposium on *Nocardia* and *Streptomyces,* M. Mordarski, W. Kurylowicz and J. Jeljaszewicz, pp. 145–51. Gustar Fischer Verlag: Stuttgart.

Orndorff, S.A. and Colwell, R.R. (1980). Distribution and identification of luminous bacteria from the Sargasso Sea. *Applied and Environmental Microbiology,* **39**, 983–7.

Patrick, F.M. (1978). The use of membrane filtration and marine agar 2216E to enumerate marine heterotrophic bacteria. *Aquaculture,* **13**, 369–72.

Rassoulzadean, F. and Sheldon, R.W. (1986). Predator–prey interactions of nano-plankton and bacteria in an oligotrophic marine environment. *Limnology and Oceanography,* **31**, 1010–21.

Ruby, E.G. and Nealson, K.H. (1978). Seasonal changes in the species composition of luminous bacteria in nearshore seawater. *Limnology and Oceanography,* 23, 530–3.

Seliger, H.H., Carpenter, J.H., Loftus, M. and McElroy, W.D. (1970). Mechanisms for the accumulation of high concentrations of dinoflagellates in a bioluminescent bay. *Limnology and Oceanography,* **15**, 234–45.

Sherr, B. and Sherr, E. (1983). Enumeration of heterotrophic microprotozoa by epifluorescence microscopy. *Estuarine and Coastal Shelf Science,* **16**, 1–7.

Sherr, E.B., Sherr, B.F., Fallon, R.D. and Newell, S.Y. (1986). Small aloricate ciliates as a major component of the marine heterotrophic nanoplankton. *Limnology and Oceanography,* **31**, 177–83.

Sieburth, J.McN. (1971). Distribution and activity of oceanic bacteria. *Deep Sea Research,* **18**, 1111–21.

Sieburth, J.McN. (1979). *Sea Microbes.* Oxford University Press: New York.

Sieburth, J.McN., Willis, P.J., Johnson, K.M., Burney, C.M., Lavoie, D.M., Hinga, K.R., Caron, D.A., French, F.W., Johnson, P.W. and Davis, P.G. (1976). Dissolved organic matter and heterotrophic microneuston in the surface microlayers of the North Atlantic. *Science (Washington, DC),* **194**, 1415–18.

Simidu, U., Lee, W.J. and Kogure, K. (1983). Comparison of different techniques for determining plate counts of marine bacteria. *Bulletin of the Japanese Society of Scientific Fisheries,* **49**, 1199–203.

Sleigh, M.A. (1973). *The Biology of Protozoa.* Edward Arnold: London.

Sorokin, Ju. I. (1971). Abundance and production of bacteria in the open water of the Central Pacific. *Oceanology,* **11**, 85–94.

Staley, J.T. and Konopka, A. (1985). Measurement of *in situ* activities of nonphotosynthetic microorganisms in aquatic and terrestrial habitats. *Annual Review of Microbiology,* **39**, 321–46.

Sugahara, I., Sugiyama, M. and Kawai, A. (1974). Distribution and activity of nitrogen-cycle bacteria in water-sediment systems with different concentrations of oxygen. In *Effect of the Ocean Environment on Microbial Activities,* R.R. Colwell and R.Y. Morita, pp. 327–40. University Park Press; Baltimore.

Taga, N. and Matsuda, O. (1974). Bacterial populations attached to plankton and detritus in seawater. In *Effect of the Ocean Environment on Microbial Activities,* R.R. Colwell and R.Y. Morita, pp. 433–48. University Park Press: Baltimore.

Takahashi, M., Kikuchi, K. and Hara, Y. (1985). Importance of picocyanobacteria biomass (unicellular, blue-green algae) in the phytoplankton population of the coastal waters off Japan. *Marine Biology,* **89**, 63–9.

Taylor, V.I., Baumann, P., Reichelt, J.L. and Allen, R.D. (1974). Isolation, enumeration, and host range of marine bdellovibrios. *Archives of Microbiology*, **98**, 101–14.

Throndsen, J. (1969). Flagellates of Norwegian coastal water. *Nyatt Magasin foer Botanikk (Oslo)*, **16**, 161–216.

Tilton, R.C., Cobet, A.B. and Jones, G.E. (1967). Marine thiobacilli. I. Isolation and distribution. *Canadian Journal of Microbiology*, **13**, 1329–34.

Travers, A. and Travers, M. (1968). Les Silicoflagellates du Golfe de Marseille. *Marine Biology*, **1**, 285–8.

Tuttle, J.H. and Jannasch, H.W. (1972). Occurrence and types of *Thiobacillus*–like bacteria in the sea. *Limnology and Oceanography*, **17**, 532–43.

Väätänen, P. (1980). Effects of environmental factors on microbial populations in brackish water off the Southern coast of Finland. *Applied and Environmental Microbiology*, **40**, 48–54.

Velji, M.I. and Albright, L.J. (1986). Microscopic enumeration of attached marine bacteria of sea water, marine sediment, fecal matter, and kelp blade samples following pyrophosphate and ultrasound treatments. *Canadian Journal of Microbiology*, **32**, 121–6.

Waterbury, J.B. and Stanier, R.Y. (1981). Chapter 9. Isolation and growth of cyanobacteria from marine and hypersaline environments. In *The Prokaryotes. A Handbook on Habitats, Isolation and Identification of Bacteria*, M.P. Starr, H. Stolp, H.G. Trüper, A. Balows and H.G. Schlegel, (eds), vol. 1, pp. 221–3. Springer-Verlag: Berlin.

Waterbury, J.B., Watson, S.W., Guillard, R.R. and Brand, L.E. (1979). Widespread occurrence of a unicellular, marine, planktonic cyanobacterium. *Nature (London)*, **277**, 293–294.

Watson, E.T. and Williams, S.T. (1974). Studies on the ecology of actinomycetes in soil. VII. Actinomycetes in a coastal sand belt. *Soil Biology and Biochemistry*, **6**, 43–52.

Watson, S.W., Novitsky, T.J., Quinby, H.L. and Valois, F.W. (1977). Determination of bacterial number and biomass in the marine environment. *Applied and Environmental Microbiology*, **33**, 940–6.

Weber, F.H. and Greenberg, E.P. (1981). Rifampin as a selective agent for the enumeration and isolation of spirochetes from salt marsh habitats. *Current Microbiology*, **5**, 303–6.

Weibe, W.J. and Hendricks, C.W. (1974). Distribution of heterotrophic bacteria in a transect of the Antarctic Ocean. In *Effect of the Ocean Environment on Microbial Activities*, R.R. Colwell and R.Y. Morita (eds), pp. 524–35. University Park Press: Baltimore.

Weise, W. and Rheinheimer, G. (1978). Scanning electron microscopy and epifluorescence investigation of bacterial colonization of marine sand sediments. *Microbial Ecology*, **4**, 175–88.

Westlake, D.W.S., Belicek, W. and Cook, F.D. (1976). Microbial utilization of raw and hydrogenated shale oils. *Canadian Journal of Microbiology*, **22**, 221–7.

Weyland, H. (1969). Actinomycetes in North Sea and Atlantic Ocean sediments. *Nature (London)*, **223**, 858.

Wilson, C.A. and Stevenson, L.H. (1980). The dynamics of the bacterial population associated with a salt marsh. *Journal of Experimental Marine Biology and Ecology*, **48**, 123–38.

Wright, R.T. and Coffin, R.B. (1984). Measuring microzooplankton grazing on planktonic marine bacteria by its impact on bacterial production. *Microbial Ecology*, **10**, 137–49.

Yamamoto, N. and Lopez, G. (1985). Bacterial abundance in relation to surface area and organic content of marine sediments. *Journal of Experimental Marine Biology and Ecology*, **90**, 209–20.

Yanagita,T., Ichikawa,T.,Tsuji,T., Kamata,Y., Ito, K. and Sasaki, M. (1978).Two trophic groups of bacteria, oligotrophs and eutrophs: their distributions in fresh and sea water areas in the central northern Japan. *Journal of General and Applied Microbiology*, **24**, 59–88.

Zevenboom,W. (1986).Tracing red-pigmented marine cyanobacteria using *in vivo* absorption maxima. *FEMS Microbiology Ecology*, **38**, 267–75.

Chapter 4

Anderson, J.I.W. (1962). Studies on micrococci isolated from the North Sea. *Journal of Applied Bacteriology*, **25**, 362–8.

Austin, B. (1982).Taxonomy of bacteria isolated from a coastal, marine fish-rearing unit. *Journal of Applied Bacteriology*, **53**, 253–68.

Balch, W.E., Fox, G.E., Magnum, L.T., Woese, C.R. and Wolfe, R.S. (1979). Methanogens: re-evaluation of a unique biological group. *Microbiological Reviews*, **43**, 260–96.

Bauld, J. and Staley, J.T. (1976). *Planctomyces maris* sp. nov.: a marine isolate of the *Planctomyces/Blastocaulis* group of budding bacteria. *Journal of General Microbiology*, **97**, 45–55.

Bauld, J. and Staley, J.T. (1980). *Planctomyces maris* sp. nov.: nom. rev. *International Journal of Systematic Bacteriology*, **30**, 657.

Bauld, J., Bigford, R. and Staley, J.T. (1983). *Prosthecomicrobium litoralum*, a new species from marine habitats. *International Journal of Systematic Bacteriology*, **33**, 613–7.

Baumann, P. and Baumann, L. (1984). Genus 11. *Photobacterium* Beijerinck 1889, 401[AL]. In *Bergey's Manual of Systematic Bacteriology*, N.R. Krieg (ed), vol. 1, pp. 539–45. Williams & Wilkins: Baltimore.

Baumann, P., Baumann, L., Bowditch, R. D. and Beaman, B. (1984a).Taxonomy of *Alteromonas: A. nigrifaciens* sp. nov., nom. rev.; *A. macleodii;* and *A. haloplanktis. International Journal of Systematic Bacteriology*, **34**, 145–9.

Baumann, L., Bowditch, R. and Baumann, P. (1983). Description of *Deleya* gen. nov. created to accommodate the marine species *Alcaligenes aestus, A. pacificus, A. cupidus, A. venustus*, and *Pseudomonas marina. International Journal of Systematic Bacteriology*, **33**, 793–802.

Baumann, P., Furniss, A.L. and Lee, J.V. (1984b). Genus 1. *Vibrio* Pacini 1854, 411[AL]. In *Bergey's Manual of Systematic Bacteriology*, N.R. Krieg (ed), vol. 1., pp. 518–38. Williams & Wilkins: Baltimore.

Boeyé, A. and Aerts, M. (1976). Numerical taxonomy of *Bacillus* isolates from North Sea sediments. *International Journal of Systematic Bacteriology*, **26**, 427–41.

Bouvet, P.J.M. and Grimont, P.A.D. (1986).Taxonomy of the genus *Acinetobacter* with the recognition of *Acinetobacter baumannii* sp. nov., *Acinetobacter haemolyticus* sp. nov., *Acinetobacter johnsonii* sp. nov., and *Acinetobacter junii* sp. nov. and emended descriptions of *Acinetobacter calcoaceticus* and *Acinetobacter lwoffii*. *International Journal of Systematic Bacteriology*, **36**, 228–40.

Brock, T.D. (1981). Chapter 24. The genus *Leucothrix*. In *The Prokaryotes A Handbook on Habitats, Isolation, and Identification of Bacteria*, M.P. Starr, H. Stolp, H.G. Trüper, A. Balows and H.G. Schlegel (eds), vol. 1, pp. 400–8. Springer Verlag: Berlin.

Burnham, J.C. and Conti, S.F. (1984). Genus *Bdellovibrio* Stolp and Starr 1963, 243[AL]. In *Bergey's Manual of Systematic Bacteriology*, N.R. Krieg (ed), vol. 1, pp. 118–24. Williams & Wilkins: Baltimore.

Canale-Parola, E. (1984). Genus 1. *Spirochaeta* Ehrenberg 1835, 313[AL]. In *Bergey's Manual of Systematic Bacteriology*, N.R. Krieg (ed), vol. 1, pp. 39–46. Williams & Wilkins: Baltimore.

Cross, T. (1981). Chapter 157. The monosporic actinomycetes. In *The Prokaryotes, A Handbook on Habitats, Isolation, and Identification of Bacteria*, M.P. Starr, H. Stolp, H.G. Trüper, A. Balows and H.G. Schlegel (eds), vol. II, pp. 2091–102. Springer Verlag: Berlin.

Dodge, J.D. (1982). *Marine Dinoflagellates of the British Isles*. H.M.S.O.: London.

Drews, G. (1981). *Rhodospirillium salexigens*, spec. nov., an obligatory halophilic phototrophic bacterium. *Archives of Microbiology*, **130**, 325–7.

Florenzano, G., Balloni, W. and Materassi, R. (1986). Nomenclature of *Prochloron didemni* (Lewin 1977) sp. nov., nom. rev., *Prochloron* (Lewin 1976) gen. nov., nom. rev., *Prochloraceae* fam. nov., *Prochlorales* ord. nov., nom. rev. in the class *Photobacteria* Gibbons and Murray. *International Journal of Systematic Bacteriology*, **36**, 351–3.

Gibson, J., Pfennig, N. and Waterbury, J.B. (1984). *Chloroherpeton thalassium* gen. nov. et spec. nov., a non-filamentous, flexing and gliding green sulfur bacterium. *Archives of Microbiology*, **138**, 96–101.

Goodfellow, M. and Minnikin, D.E. (1981). Chapter 155. The genera *Nocardia* and *Rhodococcus*. In *The Prokaryotes. A Handbook on Habitats, Isolation and Identification of Bacteria*, M.P. Starr, H. Stolp, H.G. Trüper, A. Balows and H. G. Schlegel (eds), vol. II, pp. 2016–27. Springer Verlag: Berlin.

Gottschalk, G., Andreesen, J.R. and Hippe, H. (1981). Chapter 138. The genus *Clostridium* (nonmedical aspects). In *The Prokaryotes. A Handbook on Habitats, Isolation and Identification of Bacteria*, M.P. Starr, H. Stolp, H.G. Trüper, A. Balows and H.G. Schlegel (eds), vol. II, pp. 1767–803. Springer Verlag: Berlin.

Grimont, P.A.D. and Grimont, F. (1984). Genus VIII. *Serratia* Bizio 1823, 288. In *Bergey's Manual of Systematic Bacteriology*, N.R. Krieg (ed), vol. I, pp. 477–84. Williams & Wilkins: Baltimore.

Guelin, A., Michustina, I.E., Goulevskaya, S.A., Petchnikov, N.V. and Ledoya, L.A. (1977). Étude sur les microvibrions marin de Roscoff (*Microvibrio marinus roscoffensis*). *Comptes Rendus de Academie Sciences (Paris)*, **284**, 2171–4.

Gunn, B.A. and Colwell, R.R. (1983). Numerical taxonomy of staphylococci isolated from the marine environment. *International Journal of Systematic Bacteriology*, **33**, 751–9.

Hamilton, R.D. and Austin, K.E. (1967). Physiological and cultural characteristics of *Chromobacterium marinum* sp. n. *Antonie van Leeuwenhoek*, **33**, 257–64.

Hansen, T.A. and Veldkamp, H. (1973). *Rhodopseudomonas sulfidophila*, nov. spec., a new species of the purple non-sulphur bacteria. *Archives of Microbiology*, **92**, 45–58.

Harwood, C.S. and Canale-Parola, E. (1983). *Spirochaeta isovalerica* sp. nov., a marine anaerobe that forms branched-chain fatty acids as fermentation products. *International Journal of Systematic Bacteriology*, **33**, 573–9.

Helmke, E. and Weyland, H. (1984). *Rhodococcus marinonascens* sp. nov., an actinomycete from the sea. *International Journal of Systematic Bacteriology*, **34**, 127–38.

Hendey, N.I. (1964). *An introductory account of the smaller algae of British coastal waters. Part V: Bacillariophyceae (Diatoms)*, Fisheries Investigations Series IV. H.M.S.O: London.

Holmes, B., Owen, R.J. and McMeekin, T.A. (1984). Genus *Flavobacterium* Bergey, Breed, Hammer and Huntoon 1923, 97[AL]. In *Bergey's Manual of Systematic Bacteriology*, N.R. Krieg (ed), vol. 1, pp. 353–61. Williams & Wilkins: Baltimore.

Imhoff, J.F. (1983) *Rhodopseudomonas marina* sp. nov., a new marine phototrophic purple bacterium. *Systematic and Applied Microbiology*, **4**, 512–21.

Janvier, M., Frehel, C., Grimont, F. and Gasser, F. (1985). *Methylophaga marina* gen. nov., sp. nov. and *Methylophaga thalassica* sp. nov., marine methylotrophs. *International Journal of Systematic Bacteriology*, **35**, 131–9.

Jones, W.J., Paynter, M.J.B. and Gupta, R. (1983). Characterization of *Methanococcus maripaludis* sp. nov., a new methanogen isolated from salt marsh sediment. *Archives of Microbiology*, **135**, 91–7.

Juni, E. (1984). Genus III. *Acinetobacter* Brisou and Prévot 1954, 727[AL]. In *Bergey's Manual of Systematic Bacteriology* N.R. Krieg (ed), vol. 1, pp. 303–7. Williams & Wilkins: Baltimore.

Kersters, K. and De Ley, J. (1984). Genus *Alcaligenes* Castellani and Chalmers 1919, 936[AL]. In *Bergey's Manual of Systematic Bacteriology*, N.R. Krieg (ed), vol. 1, pp. 361–73. Williams & Wilkins: Baltimore.

Kocur, M. (1984). Genus *Paracoccus* Davis 1969, 384[AL]. In *Bergey's Manual of Systematic Bacteriology*, N.R. Krieg (ed), vol. 1, pp. 399–402. Williams & Wilkins: Baltimore.

Kohlmeyer, J. (1986). Taxonomic studies of the marine Ascomycotina. In *The Biology of Marine Fungi*, S.T. Moss (ed), pp. 199–210. Cambridge University Press: Cambridge.

Kohlmeyer, J. and Kohlmeyer, E. (1979). *Marine Mycology: the Higher Fungi*. Academic Press: New York.

Krieg, N.R. (1984). Genus *Oceanospirillum* Hylemon, Wells, Krieg and Jannasch 1973, 361[AL]. In *Bergey's Manual of Systematic Bacteriology*, N.R. Krieg (ed), vol. 1, pp. 104–10. Williams & Wilkins: Baltimore.

Kubica, G.P. and Good, R.C. (1981). Chapter 150. The genus *Mycobacterium* (except *M. leprae*). In *The Prokaryotes A Handbook on Habitats, Isolation and Identification of Bacteria*, M.P. Starr, H. Stolp, H.G. Trüper, A. Balows and H.G. Schlegel (eds), vol. II, pp. 1962–84. Springer Verlag: Berlin.

Kuenen, J.G. and Tuovinen, O.H. (1981). Chapter 81. The genera *Thiobacillus* and *Thiomicrospira*. In *The Prokaryotes A Handbook on Habitats, Isolation and*

Identification of Bacteria M.P. Starr, H. Stolp, H.G.Trüper, A. Balows and H.G. Schlegel (eds), vol. I, pp. 1023–36. Springer Verlag: Berlin.

Kuenen, J.G. and Veldkamp, H. (1972). *Thiomicrospira pelophila* gen. n., sp. n., a new obligately chemolithotrophic colourless sulphur bacterium. *Antonie van Leeuwenhoek*, **38**, 241–56.

Kutzner, H.J. (1981). Chapter 156. The family Streptomycetaceae. In *The Prokaryotes A Handbook on Habitats, Isolation and Identification of Bacteria*, M.P. Starr, H. Stolp, H.G.Trüper, A. Balows and H.G. Schlegel (eds), vol. II, pp. 2028–90. Springer Verlag: Berlin.

Larkin, J.M. and Borrall, R. (1984). Genus III. *Flectobacillus* Larkin, Williams and Taylor 1977, 152[AL]. In *Bergey's Manual of Systematic Bacteriology*, N.R. Krieg (ed), vol. 1, pp. 129–32. Williams & Wilkins: Baltimore.

La Rivière, J.W.M. and Schmidt, K. (1981). Chapter 82. Morphologically conspicious sulphur-oxidising eubacteria. In *The Prokaryotes A Handbook on Habitats, Isolation and Identification of Bacteria*, M.P. Starr, H. Stolp, H.G. Trüper, A. Balows and H.G. Schlegel (eds), vol. I, pp. 1037–48. Springer Verlag: Berlin.

Larsen, H. (1984). Genus I. *Halobacterium* Elazari-Volcani 1957, 207[AL]. In *Bergey's Manual of Systematic Bacteriology* N.R. Krieg (ed), vol. 1, pp. 262–6. Williams & Wilkins: Baltimore.

Lewin, R.A. (1976). Prochlorophyta as a proposed new division of algae. *Nature (London)*, **261**, 697–8.

Lewin, R.A. (1977). *Prochloron*, type genus of the Prochlorophyta. *Phycologia*, **16**, 217.

MacDonell, M.T. and Colwell, R.R. (1985). Phylogeny of the *Vibrionaceae*, and recommendation for two new genera, *Listonella* and *Shewanella*. *Systematic and Applied Microbiology*, **6**, 171–82.

Maer, S. and Gallardo, V.A. (1984). *Thioploca araucae* sp. nov. and *Thioploca chileae* sp. nov. *International Journal of Systematic Bacteriology*, **34**, 414–8.

Maier, S. (1974). Genus III. *Thioploca* Lauterborn. In *Bergey's Manual of Determinative Bacteriology*, 8th edn, R.E. Buchanan and N.E. Gibbons (eds), pp. 115–6. Williams & Wilkins: Baltimore.

McIntire, C.D. and Moore, W.W. (1977). Chapter II. Marine littoral diatoms: ecological considerations. In *The Biology of Diatoms*, D. Werner (ed), pp. 333–71. Botanical Monographs No. 13. Blackwells: Oxford.

Meglitsch, P.A. (1972). Chapter 3. Protozoa: acellular animals. In *Invertebrate Zoology*, 2nd edn., pp. 16–77. Oxford University Press: New York.

Moore, R.L. (1981). The biology of *Hyphomicrobium* and other prosthecate, budding bacteria. *Annual Review of Microbiology*, **35**, 567–94.

Nesterenko, G.A., Nognia, T.M., Kasumova, S.A., Kvasnikov, E.J. and Batrakov, S.G. (1982). *Rhodococcus luteus* nom. nov. and *Rhodococcus maris* nom. nov. *International Journal of Systematic Bacteriology*, **32**, 1–14.

Nishimura, Y., Kanbe, K. and Iizuka, H. (1986). Taxonomic studies of aerobic coccobacilli from seawater. *Journal of General and Applied Microbiology*, **32**, 1–11.

Oren, A., Weisburg, W.G., Kessel, M. and Woese, C.R. (1984). *Halobacteroides halobius* gen. nov., sp. nov., a moderately halophilic anaerobic bacterium from the bottom sediments of the Dead Sea. *Systematic and Applied Microbiology*, **5**, 8–70.

Pfennig, N. (1984). Genus *Desulfuromonas* Pfennig and Biebl 1977, 306^{AL}. In *Bergey's Manual of Systematic Bacteriology*, N.R. Krieg (ed), vol. 1, pp. 664–6. Williams & Wilkins: Baltimore.

Poindexter, J.S. (1964). Biological properties and classification of the *Caulobacter* group. *Bacteriological Reviews*, **28**, 231–95.

Postgate, J.R. (1984). Genus *Desulfovibrio* Kluyver and van Niel 1936, 397^{AL}. In *Bergey's Manual of Systematic Bacteriology*, N.R. Krieg (ed), vol. 1, pp. 666–72. Williams & Wilkins: Baltimore.

Reichenbach, H. and Dworkin, M. (1981). Chapter 21. The order Cytophagales (with addenda on the genera *Herpetosiphon, Saprospira*, and *Flexithrix*). In *The Prokaryotes A Handbook on Habitats, Isolation and Identification of Bacteria*, M.P. Starr, H. Stolp, H.G. Trüper, A. Balows and H.G. Schlegel (eds), vol. I, pp. 356–79. Springer Verlag: Berlin.

Rippka, R., Waterbury, J.B. and Stanier, R.Y. (1981). Chapter 12. Provisional generic assignments for cyanobacteria in pure culture. In *The Prokaryotes A Handbook on Habitats, Isolation and Identification of Bacteria*, M.P. Starr, H. Stolp, H.G. Trüper, A. Balows and H.G. Schlegel (eds), vol. I, pp. 247–56. Springer Verlag: Berlin.

Rivard, C.J. and Smith, P.H. (1982). Isolation and characterization of a thermophilic marine methanogenic bacterium, *Methanogenium thermophilicum* sp. nov. *International Journal of Systematic Bacteriology*, **32**, 430–6.

Romesser, J.A., Wolfe, R.S., Mayer, F., Spiess, E. and Walther-Mauruschat, A. (1979). *Methanogenium*, a new genus of marine methanogenic bacteria and characterization of *Methanogenium cariaci* sp. nov. and *Methanogenium marisnigri* sp. nov. *Archives of Microbiology*, **121**, 147–53.

Rüger, H.–J. (1983). Differentiation of *Bacillus globisporus, Bacillus marinus* comb. nov., *Bacillus aminovorans*, and *Bacillus insolitus. International Journal of Systematic Bacteriology*, **33**, 157–61.

Schaal, K.P. and Pulverer, G. (1981). Chapter 148. The genera *Actinomyces, Agromyces, Arachnia, Bacterionema* and *Rothia*. In *The Prokaryotes A Handbook on Habitats, Isolation and Identification of Bacteria*, M.P. Starr, H. Stolp, H.G. Trüper, A. Balows and H.G. Schlegel (eds), vol. II, pp. 1923–50. Springer Verlag: Berlin.

Schlesner, H. (1986). *Pirella marina* sp. nov., a budding, peptidoglycan-less bacterium from brackish water. *Systematic and Applied Microbiology*, **8**, 177–80.

Schlesner, H. and Stackebrandt, E. (1986). Assignment of the genera *Planctomyces* and *Pirella* to a new family *Planctomycetaceae* fam. nov. and description of the order *Planctomycetales* ord. nov. *Systematic and Applied Microbiology*, **8**, 174–6.

Shiba, T. and Simidu, U. (1982). *Erythrobacter longus* gen. nov., sp. nov., an aerobic bacterium which contains bacteriochlorophyll a. *International Journal of Systematic Bacteriology*, **32**, 211–17.

Skerman, V.B.D., McGowan, V. and Sneath, P.H.A. (1980). Approved lists of bacterial names. *International Journal of Systematic Bacteriology*, **30**, 225–420.

Sleigh, M. (1973). *The Biology of Protozoa*. Edward Arnold: London.

Smith, L.D.S. (1970). *Clostridium oceanicum*, sp. n., a spore-forming anaerobe isolated from marine sediments. *Journal of Bacteriology*, **103**, 811–3.

Sowers, K.R. and Ferry, J.G. (1983). Isolation and characterization of a methylotrophic marine methanogen, *Methanococcoides methylutenis* gen. nov., sp. nov. *Applied and Environmental Microbiology*, **45**, 684–90.

Stanier, R.Y. (1974). Division 1. The cyanobacteria. In *Bergey's Manual of Determinative Bacteriology*, 8th edn, R.E. Buchanan and N.E. Gibbons (eds), p. 22. Williams & Wilkins: Baltimore.

Trüper, H.G. and Imhoff, J.F. (1981). Chapter 15. The genus *Ectothiorhodospira*. In *The Prokaryotes A Handbook on Habitats, Isolation and Identification of Bacteria*, M.P. Starr, H. Stolp, H.G. Trüper, A. Balows and H.G. Schlegel (eds), vol. I, pp. 274–8. Springer Verlag: Berlin.

Trüper, H.G. and Pfennig, N. (1981). Chapter 18. Characterization and identification of the anoxygenic phototrophic bacteria. In *The Prokaryotes A Handbook on Habitats, Isolation and Identification of Bacteria*, M.P. Starr, H. Stolp, H.G. Trüper, A. Balows and H.G. Schlegel (eds), vol. I, pp. 299–312. Springer Verlag: Berlin.

Van Hao, M., Kocur, M. and Komagata, K. (1984). *Marinococcus* gen. nov., a new genus for motile cocci with meso-diaminopimelic acid in the cell wall; and *Marinococcus albus* sp. nov. and *Marinococcus halophilus* (Novitsky and Kushner) comb. nov. *Journal of General and Applied Microbiology*, 30, 449–59.

Van Landschoot, A. and De Ley, J. (1983). Intra- and intergeneric similarities of the rRNA cistrons of *Alteromonas*, *Marinomonas* (Gen. nov.) and some other Gram-negative bacteria. *Journal of General Microbiology*, **129**, 3057–74.

Vreeland, R.H. (1984). Genus *Halomonas* Vreeland, Litchfield, Martin and Elliot 1980, 494[VP]. In *Bergey's Manual of Systematic Bacteriology*, N.R. Krieg (ed), vol. 1, pp. 340–3. Williams & Wilkins: Baltimore.

Walsby, A.E. (1981). Chapter 10. Cyanobacteria: planktonic gas-vacuolate forms. In *The Prokaryotes A Handbook on Habitats, Isolation and Identification of Bacteria*, M.P. Starr, H. Stolp, H.G. Trüper, A. Balows and H.G. Schlegel (eds), vol. I, pp. 224–35. Springer Verlag: Berlin.

Waterbury, J.B. and Stanier, R.Y. (1981). Chapter 9. Isolation and growth of cyanobacteria from marine and hypersaline environments. In *The Prokaryotes A Handbook on Habitats, Isolation and Identification of Bacteria*, M.P. Starr, H. Stolp, H.G. Trüper, A. Balows and H.G. Schlegel (eds), vol. I, pp. 221–3. Springer Verlag: Berlin.

Watson, St.W., Valois, F.W. and Waterbury, J.B. (1981). Chapter 80. The family Nitrobacteraceae. In *The Prokaryotes A Handbook on Habitats, Isolation and Identification of Bacteria*, M.P. Starr, H. Stolp, H.G. Trüper, A. Balows and H.G. Schlegel (eds), vol. I, pp. 1005–22. Springer Verlag: Berlin.

Watson, St.W., Bock, E., Valois, F.W., Waterbury, J.B. and Schlosser, U. (1986). *Nitrospira marina* gen nov., sp. nov.: a chemolithotrophic nitrite-oxidising bacterium. *Archives of Microbiology*, **144**, 1–7.

Weiner, R.M., Devine, R.A., Powell, D.M., Dagasan, L. and Moore, R.L. (1985). *Hyphomonas oceanitis* sp. nov., *Hyphomonas hirschiana* sp. nov., and *Hyphomonas jannaschiana* sp. nov. *International Journal of Systematic Bacteriology*, **35**, 237–43.

Weyland, H. (1969). Actinomycetes in North Sea and Atlantic Ocean sediments. *Nature (London)*, **223**, 858.

Weyland, H. (1981a). Distribution of actinomycetes on the sea floor. *Zentralblatt für Bakteriologie, Parasitenkunde, Infektionskrankheiten und Hygiene, Abteilung I Supplement II*, 185–93.

Weyland, H. (1981b). Characteristics of actinomycetes isolated from marine sediments. *Zentralblatt für Bakteriologie, Parasitenkunde, Infektionskrankheiten und Hygiene, Abteilung I Supplement II*, 309–14.

Weyland, H., Rüger, H.-J. and Schwarz, H. (1970). Zur Isolierung und Identifizierung mariner Bakterien. *Veroefflichungen der Instituts für Meeresforschung in Bremerhaven*, **12**, 269–96.

Widdel, F. and Pfennig, N. (1984a). Genus *Desulfobacter* Widdel 1981, 382^VP (effective publication: Widdel 1980, 373). In *Bergey's Manual of Systematic Bacteriology*, N.R. Krieg (ed), vol. 1, pp. 674–6. Williams & Wilkins: Baltimore.

Widdel, F. and Pfennig, N. (1984b). Genus *Desulfobulbus* Widdel 1981, 382^VP (effective publication: Widdel 1980, 374). In *Bergey's Manual of Systematic Bacteriology*, N.R. Krieg (ed), vol. 1, pp. 676–7. Williams & Wilkins: Baltimore.

Widdel, F. and Pfennig, N. (1984c). Genus *Desulfococcus* Widdel 1981, 382^VP (effective publication: Widdel 1980, 376). In *Bergey's Manual of Systematic Bacteriology*, N.R. Krieg (ed), vol. 1, pp. 673–4. Williams & Wilkins: Baltimore.

Widdel, F. and Pfennig, N. (1984d). Genus *Desulfosarcina* Widdel 1981, 382^VP (effective publication: Widdel 1980, 382). In *Bergey's Manual of Systematic Bacteriology*, N.R. Krieg (ed), vol. 1, pp. 677–9. Williams & Wilkins: Baltimore.

Wiessner, W. (1981). Chapter 22. The family Beggiatoaceae. In *The Prokaryotes A Handbook on Habitats, Isolation and Identification of Bacteria*, M.P. Starr, H. Stolp, H.G. Trüper, A. Balows and H.G. Schlegel (eds), vol. I, pp. 380–9. Springer Verlag: Berlin.

Wildgruber, G., Thomm, M., König, H., Ober, K., Ricchiuto, T. and Stetter, K.O. (1982). *Methanoplanus limicola*, a plate-shaped methanogen representing a novel family, the Methanoplanaceae. *Archives of Microbiology*, **132**, 31–6.

Zobell, C.E. and Upham, H.C. (1944). A list of marine bacteria including descriptions of sixty new species. *Bulletin of the Scripps Institution of Oceanography, Technical Series*, **5**, 239–92.

Chapter 5

Alldredge, A.L., Cole, J.J. and Caron, D.A. (1986). Production of heterotrophic bacteria inhabiting macroscopic organic aggregates (marine snow) from surface waters. *Limnology and Oceanography*, **31**, 68–78.

Alongi, D.M. and Tietjen, J.H. (1980). Population growth and trophic interactions among free-living marine nematodes. In *Marine Benthic Dynamics*, K.R. Tenore, and B.C. Coull (eds), pp. 151–66. University of South Carolina: Columbia.

Amy, P.S. and Morita, R.Y. (1983). Protein patterns of growing and starved cells of a marine *Vibrio* sp. *Applied and Environmental Microbiology*, **45**, 1748–52.

Amy, P.S., Pauling, C. and Morita, R.Y. (1983a). Starvation–survival processes of a marine vibrio. *Applied and Environmental Microbiology*, **45**, 1041–8.

Amy, P.S., Pauling, C. and Morita, R.Y. (1983b). Recovery from nutrient starvation by a marine *Vibrio* sp. *Applied and Environmental Microbiology*, **45**, 1685–90.

Andersen, P. and Sørensen, H.M. (1986). Population dynamics and trophic coupling in pelagic micro-organisms in eutrophic coastal waters. *Marine Ecology–Progress Series*, **33**, 99–109.

Anderson, J.I.W. and Heffernan, W.P. (1965). Isolation and characterization of filterable marine bacteria. *Journal of Bacteriology*, **90**, 1713–8.

Andersson, A., Lee, C., Azam, F. and Hagström, Å. (1985). Release of amino acids and inorganic nutrients by heterotrophic marine microflagellates. *Marine Ecology–Progress Series*, **23**, 99–106.

Araki, T. and Kitamikado, M. (1978). Distribution of mannan-degrading bacteria in aquatic environments. *Bulletin of the Japanese Society of Scientific Fisheries*, **44**, 1135–9.

Atlas, R.M. and Bartha, R. (1987). *Microbial Ecology Fundamentals and Applications*, 2nd edn. Benjamin/Cummings: Menlo Park, California.

Azam, F. and Ammerman, J.W. (1984). Cycling of organic matter by bacterioplankton in pelagic marine ecosystems: microenvironmental considerations. In *Flow of Energy and Materials in Marine Ecosystems – Theory and Practice*, M.R. Fasham (ed), pp. 345–60. Plenum Press: London.

Azam, F., Fenchel, T., Field, J.G., Gray, J.S., Meyer-Reil, L.A. and Thingstad, F. (1983). The ecological role of water-column microbes in the sea. *Marine Ecology – Progress Series*, **10**, 257–63.

Berman, T., Azov, Y. and Townsend, D. (1984). Understanding oligotrophic oceans: can the eastern Mediterranean be a useful model? In *Marine Phytoplankton and Productivity: Lecture Notes on Coastal and Estuarine Studies*, O. Holm-Hansen, L. Bolis and R. Gilles (eds), vol. 8, pp. 101–12. Springer Verlag: Berlin.

Bienfang, P.K. and Szyper, J.P. (1981). Phytoplankton dynamics in the subtropical Pacific Ocean off Hawaii. *Deep Sea Research*, **28**, 981–1000.

Bienfang, P.K. and Takahashi, M. (1983). Ultraplankton growth rates in a subtropical ecosystem. *Marine Biology*, **76**, 213–8.

Blotevogel, K.H., Fischer, U. and Lüpkes, K.H. (1986). *Methanococcus frisius* sp. nov., a new methylotrophic marine methanogen. *Canadian Journal of Microbiology*, **32**, 127–31.

Bright, J.J. and Fletcher, M. (1983). Amino acid assimilation and respiration by attached and free-living populations of a marine *Pseudomonas* sp. *Microbial Ecology*, **9**, 215–26.

Brown, C.M. (1987). Nitrate metabolism by aquatic bacteria. In *Methods in Aquatic Bacteriology*. B. Austin (ed), Wiley: Chichester (in press).

Bull, A.T. (1980). Biodegradation: some attitudes and strategies of microorganisms and microbiologists. In *Contemporary Microbial Ecology*, D.C. Ellwood, J.N. Hedger, M.J. Latham, J.M. Lynch and J.H. Slater (eds), pp. 107–36. Academic Press: London.

Burke, R.A., Reid, D.F., Brooks, J.M. and Lavoie, D.M. (1983). Upper water column methane geochemistry in the eastern tropical North Pacific. *Limnology and Oceanography*, **28**, 19–32.

Burney, C.M. (1986). Bacterial utilization of total in situ dissolved carbohydrate in offshore waters. *Limnology and Oceanography*, **31**, 427–31.

Campbell, L. and Carpenter, E.J. (1986). Estimating the grazing pressure of heterotrophic nanoplankton on *Synechococcus* spp. using the sea water dilution and selective inhibitor techniques. *Marine Ecology–Progress Series*, **33**, 121–9.

Campbell, R. (1981). *Mikrobielle Ökologie*. Verlag Chemie: Weinheim.

Carlucci, A.F. and Shimp, S.L. (1974). Isolation and growth of a marine bacterium in low concentrations of substrate. In *Effect of the Ocean Environment on Microbial Activities*, R.R. Colwell and R.Y. Morita, pp. 363–7. University Park Press: Baltimore.

Caron, D.A., Davis, P.G., Madin, L.P. and Sieburth, J. McN. (1982). Heterotrophic bacteria and bacteriovorous protozoa in oceanic macroaggregates. *Science (Washington, DC)*, **218**, 795–7.

Caron, D.A., Goldman, J.C. and Dennett, M.R. (1986). Effect of temperature on growth, respiration, and nutrient regeneration by an omnivorous microflagellate. *Applied and Environmental Microbiology*, **52**, 1340–7.

Chet, I. and Mitchell, R. (1976). Ecological aspects of microbial chemotactic behaviour. *Annual Review of Microbiology*, **30**, 221–39.

Christensen, D. (1984). Determination of substrate oxidized by sulfate reduction in impact cores of marine sediment. *Limnology and Oceanography*, **29**, 189–92.

Coleman, W.G. and Leive, L. (1979). Two mutations which affect the barrier function of the *Escherichia coli* K-12 outer membrane. *Journal of Bacteriology*, **139**, 899–910.

Colwell, R.R., Brayton, P.R., Grimes, D.J., Roszak, D.B., Huq, S.A. and Palmer, L.M. (1985). Viable but non-cuturable *Vibrio cholerae* and related pathogens in the environment: implications for release of genetically engineered microorganisms. *Biotechnology*, **3**, 817–20.

Compeau, G.C. and Bartha, R. (1985). Sulfate-reducing bacteria: principal methylators of mercury in anoxic estuarine sediment. *Applied and Environmental Microbiology*, **50**, 498–502.

Conover, R.J. and Francis, V. (1973). The use of radioactive isotopes to measure the transfer of materials in aquatic food chains. *Marine Biology*, **18**, 272–83.

Corpe, W.A. (1970). An acid polysaccharide produced by a primary film-forming marine bacterium. *Developments in Industrial Microbiology*, **11**, 402–12.

Costerton, J.W., Ingram, J.M. and Cheng, K.-J. (1974). Structure and function of the cell envelope of gram-negative bacteria. *Bacteriological Reviews*, **38**, 87–110.

Daro, M.H. (1978). A simplified ^{14}C method for grazing measurements on natural planktonic populations. *Helgoländer wissenschaften Meeresuntersuchungen*, **13**, 241–8.

Dawe, L.L. and Penrose, W.R. (1978). "Bactericidal" property of seawater: death or debilitation? *Applied and Environmental Microbiology*, **35**, 829–33.

Dawson, M.P., Humphrey, B. and Marshall, K.C. (1981). Adhesion: a tactic in the survival strategy of a marine vibrio during starvation. *Current Microbiology*, **6**, 195–8.

De Flaun, M.F. and Mayer, L.M. (1983). Relationships between bacteria and grain surfaces in intertidal sediments. *Limnology and Oceanography*, **28**, 873–81.

Dempsey, M.J. (1981). Marine bacterial fouling: a scanning electron microscope study. *Marine Biology*, **61**, 305–15.

Douglas, D.J. (1984). Microautoradiography-based enumeration of photosynthetic picoplankton with estimates of carbon-specific growth rates. *Marine Ecology – Progress Series*, **14**, 223–8.

Ducklow, H.W. (1983). Production and fate of bacteria in the oceans. *Bioscience*, **33**, 494–501.

Enzinger, R.M. and Cooper, R.C. (1976). Role of bacteria and protozoa in removal of *E. coli* from estuarine waters. *Applied and Environmental Microbiology*, **31**, 758–63.

Everest, S.A., Hipkin, C.R. and Syrett, P.J. (1984). The effect of phosphate and flavin adenine dinucleotide on nitrate reductase activity by some unicellular marine algae. *Journal of Experimental Marine Biology and Ecology*, **76**, 263–75.

Fenchel, T. (1982). Ecology of heterotrophic microflagellates: adaptations to heterogeneous environments. *Marine Ecology – Progress Series*, **9**, 25–33.

Fenchel, T. (1986). The ecology of heterotrophic microflagellates. *Advances in Microbial Ecology*, **9**, 57–97.

Fenchel, T. and Blackburn, T.H. (1979). *Bacteria and Mineral Cycling.* Academic Press: London.

Fletcher, M. (1976). The effects of proteins on bacterial attachment to polystyrene. *Journal of General Microbiology*, **94**, 400–4.

Fletcher, M. (1980). Adherence of marine micro-organisms to smooth surfaces. In *Receptors and Recognition*, Series B, E.H. Beachey (ed), vol. 6, pp. 354–74. Chapman and Hall: London.

Fletcher, M. and Floodgate, G.D. (1973). An electron-microscope demonstration of an acidic polysaccharide involved in the adhesion of a marine bacterium to solid surfaces. *Journal of General Microbiology*, **74**, 325–34.

Fletcher, M. and Loeb, G.I. (1979). Influence of substratum characteristics on the attachment of a marine pseudomonad to solid surfaces. *Applied and Environmental Microbiology*, **37**, 67–72.

Fogg, G.E. (1986). Picoplankton. *Proceedings of the Royal Society of London*, **B228**, 1–30.

Fuhram, J.A. and McManus, G.B. (1984). Do bacteria-sized marine eukaryotes consume significant bacterial production? *Science (Washington, DC)*, **224**, 1257–60.

Ghiorse, W.C. and Hirsch, P. (1982). Isolation and properties of ferromanganese-depositing budding bacteria from Baltic Sea ferromanganese concretions. *Applied and Environmental Microbiology*, **43**, 1464–72.

Glover, H.E., Phinney, D.A. and Yentsch, C.S. (1985a). Photosynthetic characteristics of picoplankton compared with those of larger phytoplankton populations, in various water masses in the Gulf of Maine. *Biological Oceanography*, **3**, 223–48.

Glover, H.E., Smith, A.E. and Shapiro, L. (1985b). Diurnal variations in photosynthetic rates: comparisons of ultraphytoplankton with a larger phytoplankton size fraction. *Journal of Plankton Research*, **7**, 519–35.

Glover, H.E., Keller, M.D. and Guillard, R.R.L, (1986). Light quality and oceanic ultraphytoplankters. *Nature (London)*, **319**, 142–3.

Goldman, J.C. and Caron, D.A. (1985). Experimental studies on an omnivorous microflagellate: implications for grazing and nutrient regeneration in the marine microbial food chain. *Deep Sea Research*, **32**, 899–915.

Goldman, J.C., Caron, D.A., Andersen, O.K. and Dennett, M.R. (1985). Nutrient cycling in a microflagellate food chain. 1. Nitrogen dynamics. *Marine Ecology – Progress Series*, **24**, 231–42.

Hagström, Å. (1984). Aquatic bacteria: measurements and significance of growth. In *Current Perspectives in Microbial Ecology*, M.G. Klug and C.A. Reddy (eds), pp. 495–501. American Society for Microbiology: Washington, D.C.

Hagström, Å., Ammerman, J.W., Henrichs, S. and Azam, F. (1984). Bacterioplankton growth in seawater. II. Organic matter utilization during steady-state growth in seawater cultures. *Marine Ecology – Progress Series*, **18**, 41–8.

Hanson, R.S. (1980). Ecology and diversity of methylotrophic organisms. *Advances in Applied Microbiology*, **26**, 3–39.

Herbland, A. and Le Bouteiller, A. (1981). The size distribution of phytoplankton and particulate organic matter in the equatorial Atlantic Ocean; importance of ultraseston and consequences. *Journal of Plankton Research*, **3**, 659–73.

Hermansson, M. and Marshall, K.C. (1985). Utilization of surface localized substrate by non-adhesive marine bacteria. *Microbial Ecology*, **11**, 91–105.

Hobbie, J.E. and Melillo, J.M. (1984). Comparative carbon and energy flow in ecosystems. In *Current Perspectives in Microbial Ecology*, M.J. Klug and C.A. Reddy (eds), pp. 389–93. American Society for Microbiology: Washington, DC

Iturriaga, R. and Mitchell, B.G. (1986). Chroococcoid cyanobacteria: a significant component in the food web dynamics of the open ocean. *Marine Ecology – Progress Series*, **28**, 291–7.

Jaan, A.J., Dahllöf, B. and Kjelleberg, S. (1986). Changes in protein composition of three bacterial isolates from marine waters during short periods of energy and nutrient deprivation. *Applied and Environmental Microbiology*, **52**, 1419–21.

Jacobsen, T.R. and Azam, F. (1985). Role of bacteria in copepod fecal pellet decomposition: colonization, growth rates and mineralization. *Bulletin of Marine Science*, **35**, 495–502.

Jamieson, W., Madri, P. and Claus, G. (1976). Survival of certain pathogenic micro-organisms in sea water. *Hydrobiologia*, **50**, 117–21.

Jannasch, H.W. (1967). Growth of marine bacteria at limiting concentrations of organic carbon in sea water. *Limnology and Oceanography*, **12**, 264–71.

Joint, I.R. and Pomroy, A.J. (1983). Production of picoplankton and small nano-plankton in the Celtic Sea. *Marine Biology*, **77**, 19–27.

Joint, I.R., Owens, N.J.P. and Pomroy, A.J. (1986). Seasonal production of photosynthetic picoplankton and nanoplankton in the Celtic Sea. *Marine Ecology – Progress Series*, **28**, 251–8.

Jones, K.L. and Rhodes-Roberts, M.E. (1981). The survival of marine bacteria under starvation conditions. *Journal of Applied Bacteriology*, **50**, 247–58.

Jørgensen, B.B. (1977). The sulfur cycle of a coastal marine sediment (Limfjorden, Denmark). *Limnology and Oceanography*, **22**, 814–32.

Jørgensen, B.B. (1980). Mineralization and the bacterial cycling of carbon, nitrogen and sulphur in marine sediments. In *Contemporary Microbial Ecology*, D.C. Ellwood, J.N. Hedger, M.J. Latham, J.M. Lynch and J.H. Slater (eds), pp. 239–52. Academic Press: London.

Jørgensen, B.B. (1982). Mineralization of organic matter in the sea bed – the role of sulphate reduction. *Nature (London)*, **296**, 643–5.

King, G.M. (1984a). Metabolism of trimethylamine, choline and glycine betaine by sulphate-reducing and methanogenic bacteria in marine sediments. *Applied and Environmental Microbiology*, **48**, 719–25.

King, G.M. (1984b). Utilization of hydrogen, acetate and "noncompetitive" substrates by methanogenic bacteria in marine sediments. *Geomicrobiology Journal*, **3**, 275–306.

King, G.M., Klug, M.J. and Lovley, D.R. (1983). Metabolism of acetates, methanol and methylated amines in intertidal sediments of Lowes Cove, Maine. *Applied and Environmental Microbiology*, **45**, 1848–53.

Kjelleberg, S. and Håkansson, N. (1977). Distribution of lipolytic, proteolytic, and amylolytic marine bacteria between the lipid film and the subsurface water. *Marine Biology*, **39**, 103–9.

Kjelleberg, S., Humphrey, B.A. and Marshall, K.C. (1982). Effect of interfaces on small, starved marine bacteria. *Applied and Environmental Microbiology*, **43**, 1166–72.

Kjelleberg, S., Marshall, K.C. and Hermansson, M. (1985). Oligotrophic and copiotrophic marine bacteria – observations related to attachment. *FEMS Microbiology and Ecology*, **31**, 89–96.

Kogure, K., Simidu, U. and Taga, N. (1979). A tentative direct microscopic method for counting living marine bacteria. *Canadian Journal of Microbiology*, **25**, 415–20.

Kogure, K., Simidu, U. and Taga, N. (1980). Effect of phyto- and zooplankton on the growth of marine bacteria in filtered seawater. *Bulletin of the Japanese Society of Scientific Fisheries*, **46**, 323–6.

Koike, I., Holm-Hansen, O. and Biggs, D.C. (1986). Inorganic nitrogen metabolism by Antarctic phytoplankton with special reference to ammonium cycling. *Marine Ecology – Progress Series*, **30**, 105–16.

Large, P.J. (1983). *Methylotrophy and Methanogenesis*. American Society for Microbiology: Washington, DC.

Larsson, U. and Hagström, Å. (1982). Fractionated phytoplankton primary production, exudate release and bacterial production in a Baltic eutrophication gradient. *Marine Biology*, **67**, 57–70.

Lein, A., Namsaraev, G.B., Trotsyuk, V.Y. and Ivanov, M.V. (1981). Bacterial methanogenesis in holocene sediments of the Baltic Sea. *Geomicrobiology Journal*, **2**, 299–315.

Li, W.K.W., Subba Rao, D.V., Harrison, W.G., Smith, J.C., Cullen, J.J., Irwin, B. and Platt, T. (1983). Autotrophic picoplankton in the tropical ocean. *Science (Washington, DC)*, **219**, 292–5.

Lilley, M.D., Baross, J.A. and Gordon, L.I. (1982). Dissolved hydrogen and methane in Saanich Inlet, British Columbia. *Deep Sea Research*, **29**, 1471–84.

Lotka, A.J. (1925). *Elements of Physical Biology*. Williams and Wilkins: Baltimore.

Maeda, M. and Taga, N. (1973). Deoxyribonuclease activity in seawater and sediment. *Marine Biology*, **20**, 58–63.

Maeda, M. and Taga, N. (1974). Occurrence and distribution of deoxyribonucleic acid-hydrolyzing bacteria in seawater. *Journal of Experimental Marine Biology and Ecology*, **14**, 157–69.

Mague, T.H., Friberg, E., Hughes, D.J. and Morris, I. (1980). Extracellular release of carbon by marine phytoplankton: a physiological approach. *Limnology and Oceanography*, **25**, 262–79.

Mah, R.A., Ward, D.M., Baresi, L. and Glass, T.L. (1977). Biogenesis of methane. *Annual Review of Microbiology*, **31**, 309–41.

Malmcrona-Friberg, K., Tunbid, A., Mården, P., Kjelleberg, S. and Odham, G. (1986). Chemical changes in cell envelope and poly-ß-hydroxybutyrate during short term starvation of a marine bacterial isolate. *Archives of Microbiology*, **144**, 340–5.

Mann, S., Moench, T.T. and Williams, R.J.P. (1984). A high resolution electron microscopic investigation of bacterial magnetite. Implications for crystal growth. *Proceedings of the Royal Society of London*, **B221**, 385–93.

Marshall, K.C. (1976). *Interfaces in Microbial Ecology*. Harvard University Press: Cambridge, Mass.

Marshall, K.C., Stout, R. and Mitchell, R. (1971a). Selective sorption of bacteria from seawater. *Canadian Journal of Microbiology*, **17**, 1413–16.

Marshall, K.C., Stout, R. and Mitchell, R. (1971b). Mechanism of the initial events in the sorption of marine bacteria to surfaces. *Journal of General Microbiology*, **68**, 337–48.

Mitchell, R. (1971). Role of predators in the reversal of imbalances in microbial ecosystems. *Nature (London)*, **230**, 257–8.

Mitchell, R. and Wirsen, C. (1968). Lysis of non-marine fungi by marine micro-organisms. *Journal of General Microbiology*, **52**, 335–345.

Montagna, P.A. (1984). *In situ* measurement of meiobenthic grazing rates on sediment bacteria and edaphic diatoms. *Marine Ecology – Progress Series*, **18**, 119–30.

Moriarty, D.J.W., Iverson, R.L. and Pollard, P.C. (1986). Exudation of organic carbon by the seagrass *Halodule wrightii* Aschers, and its effect on bacterial growth in the sediment. *Journal of Experimental Marine Biology and Ecology*, **96**, 115–26.

Moriarty, D.J.W., Pollard, P.C., Hunt, W.G., Moriarty, C.M. and Wassenberg, T.J. (1985). Productivity of bacteria and micro-algae and the effect of grazing by holothurians in sediments on a coral reef flat. *Marine Biology*, **85**, 293–300.

Morita, R.Y. (1982). Starvation–survival of heterotrophs in the marine environment. *Advances in Microbial Ecology*, **6**, 171–98.

Nedwell, D.B. (1984). The input and mineralization of organic carbon in anaerobic aquatic sediments. *Advances in Microbial Ecology*, **7**, 93–131.

Nitkowski, M.F., Dudley, S. and Graikoski, J.T. (1977). Identification and characterization of lipolytic and proteolytic bacteria isolated from marine sediments. *Marine Pollution Bulletin*, **8**, 276–9.

Novitsky, J.A. (1986). Degradation of dead microbial biomass in a marine sediment. *Applied and Environmental Microbiology*, **52**, 504–9.

Novitsky, J.A. and Morita, R.Y. (1976). Morphological characterization of small cells resulting from nutrient starvation of a psychrophilic marine vibrio. *Applied and Environmental Microbiology*, **32**, 617–22.

Novitsky, J.A. and Morita, R.Y. (1977). Survival of a psychrophilic marine vibrio under long term nutrient starvation. *Applied and Environmental Microbiology*, **33**, 635–41.

Novitsky, J.A. and Morita, R.Y. (1978). Possible strategy for the survival of marine bacteria under starvation conditions. *Marine Biology*, **48**, 289–95.

Odum, E.P. (1983). *Basic Ecology*. Saunders: Philadelphia.

Olson, B.H. (1978). Enhanced accuracy of coliform testing in seawater by a modification of the Most Probable Number method. *Applied and Environmental Microbiology*, **36**, 438–44.

Oremland, R.S. (1979). Methanogenic activity in plankton samples and fish intestines: a mechanism for *in situ* methanogenesis in oceanic surface water. *Limnology and Oceanography*, **24**, 1136–41.

Paerl, H.W. (1975). Microbial attachment to particles in marine and freshwater ecosystems. *Microbial Ecology*, **2**, 73–83.

Parkes, R.J. (1987). Analysis of microbial communities within sediments using biomarkers. In *Ecology of Microbial Communities*, M. Flethcer, T.R.G. Gray and J.G. Jones (eds), pp. 147–77. Cambridge University Press: Cambridge.

Peck, H.D. and Odom, J.M. (1984). Hydrogen cycling in *Desulfovibrio*: a new mechanism for energy coupling in anaerobic microorganisms. In *Microbial Mats: Stromatolites*, Y. Cohen, R.W. Castenholz and H.O. Halvorson (eds), pp. 215–43. Allen R. Liss: New York.

Platt, T., Subba Rao, D.V., Smith, J.C., Li, W.K., Irwin, B., Horne, E.P.W. and Someoto, D.D. (1983). Photosynthetically competent phytoplankton from the aphotic zone of the deep ocean. *Marine Ecology – Progress Series*, **10**, 105–10.

Postgate, J.R. (1984). *The Sulphate-Reducing Bacteria*. Cambridge University Press: Cambridge.

Poth, M. and Focht, D.D. (1985). ^{15}N kinetic analysis of nitrous oxide production by *Nitrosomonas europaea*. *Applied and Environmental Microbiology*, **49**, 1134–41.

Probyn, T.A. (1985). Nitrogen uptake by size-fractionated phytoplankton in the southern Benguela upwelling system. *Marine Ecology – Progress Series*, **22**, 249–58.

Probyn, T.A. and Painting, S.J. (1985). Nitrogen uptake by size fractionated phytoplankton populations in Antarctic surface waters. *Limnology and Oceanography*, **30**, 1327–32.

Putt, M. and Prezelin, B. (1985). Diurnal patterns of photosynthesis in cyanobacteria and nanoplankton in California coastal waters during 'el Nino'. *Journal of Plankton Research*, **7**, 779–90.

Rieper, M. (1978). Bacteria as food for marine harpacticoid copepods. *Marine Biology*, **45**, 337–45.

Ritchie, G.A.F. and Nicholas, D.J.D. (1972). Identification of the source of nitrous oxide produced by *Nitrosomonas europaea*. *Biochemical Journal*, **126**, 1181–91.

Roman, M.R. and Rublee, P.A. (1981). A method to determine *in situ* zooplankton grazing rates on natural assemblages. *Marine Biology*. **65**, 303–9.

Roper, M.M. and Marshall, K.C. (1978). Effects of a clay mineral on microbial predation and parasitism on *Escherichia*. *Microbial Ecology*, **4**, 279–90.

Royal Society (1983). *The Nitrogen Cycle of the United Kingdom*. The Royal Society: London.

Rudd, J.W.M. and Taylor, C.D. (1980). Methane cycling in aquatic environments. *Advances in Aquatic Microbiology*, **2**, 77–150.

Saijo, Y. and Takesue, K. (1965). Further studies on the size distribution of photosynthesizing phytoplankton in the Indian Ocean. *Journal of the Oceanographic Society of Japan*, **20**, 264–71.

Sansone, F.J. and Martens, C.S. (1981). Methane production from acetate and associated methane fluxes from anoxic coastal sediments. *Science (Washington, DC)*, **211**, 707–9.

Schiemer, F. (1982). Food dependence and energetics of freeliving nematodes. I. Respiration, growth, and reproduction of *Caenorhabditis briggsae* (Nematoda) at different levels of food supply. *Oecologia*, **54**, 108–21.

Schiemer, F., Duncan, A. and Klekowski, R.C. (1980). A bioenergetic study of a benthic nematode, *Plectus palustris* de Man 1880, throughout its life cycle. II. Growth, fecundity and energy budgets at different levels of bacterial food and general ecological considerations. *Oecologia*, **44**, 205–12.

Seiler, W. and Schmidt, U. (1974). Dissolved nonconservative gases in seawater. In *The Sea*, E.D. Goldberg (ed), vol. 5, pp. 219–43. Wiley: New York.

Sherr, B.F. and Sherr, E.B. (1984). Role of heterotrophic protozoa in carbon and energy flow in aquatic ecosystems. In *Current Perspectives in Microbial Ecology*, M.J. Klug and C.A. Reddy (eds), pp. 412–23. American Society for Microbiology: Washington, DC.

Sherr, B.F., Sherr, E.B. and Berman,T. (1983). Grazing, growth, and ammonium excretion rates of a heterotrophic microflagellate fed with four species of bacteria. *Applied and Environmental Microbiology*, **45**, 1196–201.

Sieburth, J.McN. (1971). Distribution and activity of oceanic bacteria. *Deep Sea Research*, **18**, 1111–21.

Simidu, U., Lee, Won Jae and Kogure, K. (1983). Comparison of different techniques for determining plate counts of marine bacteria. *Bulletin of the Japanese Society of Scientific Fisheries*, **49**, 1199–203.

Singleton, F.L., Attwell, R.W., Jangi, M.S. and Colwell, R.R. (1982). Influence of salinity and organic nutrient concentration on survival and growth of *Vibrio cholerae* in aquatic microcosms. *Applied and Environmental Microbiology*, **43**, 1080–5.

Smith, J.C., Platt, T., Li, W.K.W., Horne, E.P.W., Harrison, W.G., Subba Rao, D.V. and Irwin, B.D. (1985). Arctic marine photoautotrophic picoplankton. *Marine Ecology – Progress Series*, **20**, 207–20.

Sørensen, J. (1982). Reduction of ferric ion in anaerobic, marine sediment and interaction with reduction of nitrate and sulfate. *Applied and Environmental Microbiology*, **43**, 319–24.

Sørensen, J., Christensen, D. and Jørgensen, B.B. (1981). Volatile fatty acids and hydrogen as substrates for sulfate-reducing bacteria in anaerobic marine sediments. *Applied and Environmental Microbiology*, **42**, 5–11.

Stal, L.J., Grossberger, S. and Krumbein, W.E. (1984). Nitrogen fixation associated with the cyanobacterial mat of a marine laminated microbial ecosystem. *Marine Biology*, **82**, 217–24.

Stevenson, L.H. (1978). A case for dormancy in aquatic systems. *Microbial Ecology*, **4**, 127–33.

Stockner, J.G. and Antia, N.J. (1986). Algal picoplankton from marine and freshwater ecosystems: a multidisciplinary perspective. *Canadian Journal of Fisheries and Aquatic Sciences*, **43**, 2472–503.

Sussman, A.S. and Halvorson, H.O. (1966). *Spores, Their Dormancy and Germination*. Harper and Row: New York.

Syrett, P.J. (1981). Nitrogen metabolism of micro-algae. In Platt, T. (ed), *Physiological Bases of Phytoplankton Ecology. Canadian Journal of Fisheries and Aquatic Sciences*, **210**, 346.

Takahashi, M. and Bienfang, P.K. (1983). Size structure of phytoplankton biomass and photosynthesis in subtropical Hawaiian waters. *Marine Biology*, **76**, 203–11.

Throndsen, J. (1978). Productivity and abundance of ultra- and nanoplankton in Oslofjorden. *Sarsia*, **63**, 273–84.

Torrella, F. and Morita, R.Y. (1981). Microcultural study of bacterial size changes and microcolony and ultramicrocolony formation by heterotrophic bacteria in seawater. *Applied and Environmental Microbiology*, **41**, 518–27.

Traganza, E.D., Swinnerton, J.W. and Cheek, C.H. (1979). Methane supersaturation and ATP–zooplankton blooms in near–surface waters of the Western Mediterranean and subtropical North Atlantic Ocean. *Deep Sea Research*, **26A**, 1237–45.

Turley, C.M. and Lochte, K. (1985). Direct measurement of bacterial productivity in stratified waters close to a front in the Irish Sea. *Marine Ecology – Progress Series*, **23**, 209–19.

Ustach, J.F. (1982). Algae, bacteria and detritus as food for the harpacticoid copepod, *Heteropsyllus pseudonunni* Coull and Palmer. *Journal of Experimental Marine Biology and Ecology*, **64**, 203–14.

Vaccaro, R.F. and Jannasch, H.W. (1966). Studies on heterotrophic activity in seawater based on glucose assimilation. *Limnology and Oceanography*, **11**, 596–607.

Volterra, V. (1926). Variazioni e fluttuazioni del numero d'individui in specie animali conviventi. *Memorial Academy Lincei*, **2**, 31–113.

Walsh, F. and Mitchell, R. (1978). Bacterial chemotactic responses in flowing water. *Microbial Ecology*, **4**, 165–8.

Ward, B.B. (1987). Kinetic studies on ammonia and methane oxidation by *Nitrosococcus oceanus*. *Archives of Microbiology*, **147**, 126–33.

Waterbury, J.B., Watson, S.W., Guillard, R.R.L. and Brand, L.E. (1979). Widespread occurrence of a unicellular, marine, planktonic cyanobacterium. *Nature (London)*, **277**, 293–4.

Watson, St.W., Bock, E., Valois, F.W., Waterbury, J.B. and Schlosser, U. (1986). *Nitrospira marina* gen. nov., sp. nov.: a chemolithotrophic nitrite-oxidizing bacterium. *Archives of Microbiology*, **144**, 1–7.

Wheeler, P.A. and Kirchman, D.L. (1986). Utilization of inorganic and organic nitrogen by bacteria in marine systems. *Limnology and Oceanography*, **31**, 998–1009.

Wikner, J., Andersson, A., Normark, S. and Hagström, Å. (1986). Use of genetically marked mini-cells as a probe in measurement of predation on bacteria in aquatic environments. *Applied and Environmental Microbiology*, **52**, 4–8.

Williams, F.M. (1980). On understanding predator–prey interactions. In *Contemporary Microbial Ecology*, D.C. Ellwood, J.N. Hedger, M.J. Latham, J.M. Lynch and J.H. Slater (eds) pp. 349–75. Academic Press: London.

Williams, P.J.LeB. (1981). Incorporation of microheterotrophic processes into the classical paradigm of the planktonic food web. *Kiel Meeresforsch*, **5**, 1–28.

Williams, P.J.LeB. and Gray, R.W. (1970). Heterotrophic utilization of dissolved organic compounds in the sea. II. Observations on the response of heterotrophic marine populations to abrupt increases in amino acid concentrations. *Journal of the Marine Biological Association of the UK*, **50**, 871–81.

Williams, P.M. (1986). Chemistry of the dissolved and particulate phases in the water column. In *Plankton Dynamics of the Southern California Bight*, R.W. Eppley (ed), pp. 53–83. Springer Verlag: Berlin.

Wilson, T.R.S. (1975). Salinity and the major elements of seawater. In *Chemical Oceanography*, J.P. Riley and G. Skirrow (eds), vol. 1, pp. 365–413. Academic Press: London.

Woodwell, G.M. (1970). The energy cycle of the biosphere. *Scientific American*, **223**, 64–74.

Wright, R.T. (1973). Some difficulties in using ^{14}C-organic solutes to measure heterotrophic bacterial activity. In *Estuarine Microbial Ecology*, L.H. Stevenson and R.R. Colwell (eds), University of South Carolina Press: Columbia.

Wright, R.T. and Coffin, R.B. (1984). Measuring microzooplankton grazing on planktonic marine bacteria by its impact on bacterial production. *Microbial*

Ecology, **10**, 137–49.

Xu, Huai-Shu, Roberts, N., Singleton, F.L., Attwell, R.W., Grimes, D.J. and Colwell, R.R. (1982). Survival and viability of nonculturable *Escherichia coli* and *Vibrio cholerae* in the estuarine and marine environment. *Microbial Ecology*, **8**, 313–23.

Zobell, C.E. and Allen, E.C. (1935). The significance of marine bacteria in the fouling of submerged surfaces. *Journal of Bacteriology*, **29**, 239–51.

Chapter 6

Actis, L.A., Potter, S.A. and Crosa, J.H. (1985). Iron-regulated outer membrane protein OM2 of *Vibrio anguillarum* is encoded by virulence plasmid pJM1. *Journal of Bacteriology*, **161**, 736–42.

Alexander, D.M. (1913). A review of piscine tubercle, with a description of an acid-fast bacillus found in the cod. *Report of the Lancashire Sea Fish Laboratory*, **21**, 43–9.

Aneer, G. and Ljungberg, O. (1976). Lymphocystis disease in Baltic herring (*Clupea harengus* var. *membras* L). *Journal of Fish Biology*, **8**, 345–50.

Apstein, R. (1910). *Cyclopterus lumpus*, der Seehases seine Fischerei und sein Mageninhalt. *Mitteilunger der Deutschen Seefischerie Versuchsanstalt*, *26*, 450–65.

Aronson, J.D. (1926). Spontaneous tuberculosis in salt water fish. *Journal of Infectious Diseases*, **39**, 315–20.

Atlas, R.M., Busdosh, M., Krichevsky, E.J. and Kaneko, T. (1982). Bacterial populations associated with the Arctic amphiphod *Boeckosimus affinis*. *Canadian Journal of Microbiology*, **28**, 92–9.

Austin, B. (1982). Taxonomy of bacteria isolated from a coastal, fish-rearing unit. *Journal of Applied Bacteriology*, **53**, 253–68.

Austin, B. (1983). Bacterial microflora associated with a coastal, marine fish-rearing unit. *Journal of the Marine Biological Association (UK)*, **63**, 585–92.

Austin, B. and Allen, D.A. (1982). Microbiology of laboratory-hatched brine shrimp (*Artemia*). *Aquaculture*, **26**, 369–83.

Austin, B. and Austin, D.A. (1986). *Bacterial Pathogens of Fish: Disease of Farmed and Wild Fish*. Ellis Horwood: Chichester.

Austin, B., Bucke, D., Feist, S.W. and Helm, M.M. (1987). Disease problems among cultured bivalve larvae. *MAFF Fisheries Publication* (in press).

Baross, J.A., Tester, P.A. and Morita, R.Y. (1978). Incidence, microscopy and etiology of exoskeleton lesions in the tanner crab, *Chionoecetes tanneri*. *Journal of the Fisheries Research Board of Canada*, **35**, 1141–9.

Birkbeck, T.H. and McHenery, J.G. (1982). Degradation of bacteria by *Mytilus edulis*. *Marine Biology*, **72**, 7–15.

Bonaveri, G.F. (1761). Quoted by Drouin de Bouville, R. de (1907). Les maladies des poissons d'eau donce d'Europe. *Annales de Sciences Agronomique*, **1**, 120–250.

Boyle, P.J. and Mitchell, R. (1978). Absence of micro-organisms in crustacean digestive tracts. *Science (Washington, DC)*, **20**, 1157–9.

Brown, C. (1981). A study of two shellfish pathogenic *Vibrio* strains isolated from a Long Island hatchery during a recent outbreak of disease. *Journal of Shellfish Research*, **1**, 83–7.

Buchanan, J.S. (1978). Cytological studies on a new species of rickettsia found in association with a phage in the digestive gland of the marine bivalve mollusc, *Tellina tenuis* (da Costa). *Journal of Fish Diseases*, **1**, 27–43.

Buchanan, J.S. and Madeley, C.R. (1978). Studies on *Herpesvirus scophthalmi* infection of turbot, *Scophthalmus maximus* (L.): Ultrastructural observations. *Journal of Fish Diseases*, **1**, 283–95.

Buchanan, J.S., Richards, R.H., Sommerville, C. and Madeley, C.R. (1978). A herpes-type virus from the turbot, *Scophthalmus maximus* (L.). *Veterinary Record*, **102**, 527–8.

Buchanan, J.S. and Richards, R.H. (1982). Herpes-type virus diseases of marine organisms. *Proceedings of the Royal Society of Edinburgh*, **81B**, 151–68.

Bullen, J.J., Roger, H.J. and Griffiths, E. (1978). Role of iron in bacterial infections. *Current Topics in Microbiology and Immunology*, **80**, 1–35.

Campbell, E.J.M., Scadding, J.G. and Roberts, R.S. (1979). The concept of disease. *British Medical Journal*, **2**, 757–62.

Cardwell, R.D. (1978). Oyster larvae mortality in south Puget Sound. *Proceedings of the National Shellfish Association*, **68**, 88–9.

Chan, E.C.S. and McManus, E.A. (1969). Distribution, characterization, and nutrition of marine micro-organisms from the algae *Polysiphonia lanosa* and *Ascophyllum nodosum*. *Canadian Journal of Microbiology*, **15**, 409–20.

Chart, H. and Trust, T.J. (1984). Characterization of the surface antigens of the marine fish pathogens, *Vibrio anguillarum* and *Vibrio ordalii*. *Canadian Journal of Microbiology*, **30**, 703–10.

Colwell, R.R. and Liston, J. (1960). Microbiology of shellfish. Bacteriological study of the natural flora of Pacific oysters (*Crassostrea gigas*). *Applied Microbiology*, **8**, 104–9.

Conover, J.T. and Sieburth, J. McN. (1964). Effect of *Sargassum* distribution on its epibiotic and antibacterial activity. *Botanica Marina*, **6**, 147–57.

Cooper, K.R., Brown, R.S. and Chang, P.W. (1982). *Journal of Invertebrate Pathology*, **39**, 149–57.

Couch, J.N. (1942). A new fungus on crab eggs. *Journal of the Elisha Mitchell Scientific Society*, **58**, 158–64.

Crosa, J.H. (1980). A plasmid associated with virulence in the marine fish pathogen *Vibrio anguillarum* specifies an iron-sequestering system. *Nature (London)*, **283**, 566–8.

Crosa, J.H. and Hodges, L.L. (1981). Outer membrane proteins induced under conditions of iron limitation in the marine fish pathogen *Vibrio anguillarum*. *Infection and Immunity*, **31**, 223–7.

Crosa, J.H., Hodges, L.L. and Schiewe, M.H. (1980). Curing of a plasmid is correlated with an attenuation of virulence in the marine fish pathogen *Vibrio anguillarum*. *Infection and Immunity*, **27**, 897–902.

Crosa, J.H., Schiewe, M.H. and Falkow, S. (1977). Evidence of plasmid contribution to the virulence of the fish pathogen *Vibrio anguillarum*. *Infection and Immunity*, **18**, 509–13.

Crosa, J.H., Walter, M.A. and Potter, S.A. (1983). The genetics of plasmid mediated virulence in the marine fish pathogen *Vibrio anguillarum*. In *Bacterial and Viral Diseases of Fish, Molecular Studies*, J.H. Crosa (ed), pp. 21–30. University of Washington: Seattle.

David, H. (1927). Über eine durch cholera ähnliche Vibrionen hervorgerufene

Fischseuche. *Zentralblatt für Bakteriologie und Parasitenkunde, Abteilung 1, Originale 102*, 46–60.

Devauchelle, G. and Vago, C. (1971). Particules d'allure virale dans les céllules del'estomac de las seiche, *Sepia officinalis* (L) (Mollusca, cephalopodes). *Comptes Rendes Academie Science de Paris*, **272**, 894–6.

DiSalvo, L.H., Blecka, J. and Zebal, R. (1978). *Vibrio anguillarum* and larval mortality in a California coastal shellfish hatchery. *Applied and Environmental Microbiology*, **35**, 219–21.

Dulin, M.P. (1979). A review of tuberculosis (mycobacteriosis) in fish. *Veterinary Medicine/Small Animal Clinician, May 1979*, 735–7.

Ellis, A.E., Waddell, I.F. and Winter, D.W. (1983). A systemic fungal disease in Atlantic salmon parr, *Salmo salar* L., caused by a species of *Phialophora. Journal of Fish Diseases*, **6**, 511–23.

Elston, R. and Leibovitz, L. (1980). Pathogenesis of experimental vibriosis in larval American oysters (*Crassostrea virginica*). *Canadian Journal of Fisheries and Aquatic Sciences*, **37**, 964–78.

Elston, R., Leibovitz, L., Relyea, D. and Zatila, J. (1981). Diagnosis of vibriosis in a commercial oyster hatchery epizootic: diagnostic tools and management features. *Aquaculture*, **24**, 53–62.

Elston, R., Elliott, E. and Colwell, R.R. (1982). Conchiolin infection and surface coating *Vibrio*; shell fragility, growth depression and mortalities in cultured oysters and clams, *Crassostrea virginica, Ostrea edulis* and *Mercenaria mercenaria. Journal of Fish Diseases*, **5**, 265–84.

Elston, R. and Lockwood, G.S. (1983). Pathogenesis of vibriosis in cultured juvenile red abalone, *Haliotis rufescens* Swainson. *Journal of Fish Diseases*, **6**, 111–28.

Evelyn, T.P.T. and Ketcheson, J.E. (1980). Laboratory and field observations on antivibriosis vaccines. In *Fish Diseases, Third COPRAQ Session*, W. Ahne (ed), pp. 45–54. Springer Verlag: Berlin.

Farley, C.A. (1976). In *Progress in Experimental Tumor Research*, vol. 20, *Tumors in Aquatic Animals*, C.J. Dawe (ed), pp. 283–94. S. Karger: Basel.

Farley, C.A., Banfield, W.G., Kasnic, G. and Foster, W.S. (1972). Oyster Herpes-type virus. *Science (New York), 1978*, 759–60.

Fisher, W.S., Nilson, E.H. and Shleser, R.A. (1975). Effect of the fungus *Haliphthoros milfordensis* on the juvenile stages of the American lobster *Homarus americanus. Journal of Invertebrate Pathology*, **26**, 41–5.

Fisher, W.S., Nilson, E.H., Steenbergen, J.F. and Lightner, D.V. (1978). Microbial diseases of cultured lobsters: a review. *Aquaculture*, **14**, 115–40.

Garland, C.D., Nash, G.V. and McMeekin, T.A. (1982). Absence of surface-associated microorganisms in adult oysters (*Crassostrea gigas*). *Applied and Environmental Microbiology*, **44**, 1205–11.

Garland, C.D., Nash, G.V., Summer, C.E. and McMeekin, T.A. (1983). Bacterial pathogens of oyster larvae (*Crassostrea gigas*) in a Tasmanian hatchery. *Australian Journal of Marine and Freshwater Research*, **34**, 483–7.

Gillespie, N.C. and Macrae, I.C. (1975). The bacterial flora of some Queensland fish and its ability to cause spoilage. *Journal of Applied Bacteriology*, **39**, 91–100.

Grabda, E. (1982). Fungi-related outgrowths of pterygophores of single fins of *Lepidopus caudatus* (Euphrasen, 1788) (Pisces:Trichiuridae). *Acta Ichthyologie et Piscat*, **12**, 87–105.

Grischkowsky, R.S. and Liston, J. (1974). Bacterial pathogenicity in laboratory-induced mortality of the Pacific oyster (*Crassostrea gigas*,Thunberg). *Proceedings of the National Shellfish Association*, **64**, 82–91.

Helm, M.M. (1971). The effect of seawater quality on the laboratory culture of *Ostrea edulis* L. larvae. *International Council for the Exploration of the Sea, C.M. Document K28*, 8p.

Hess, E. (1937). A shell disease in lobsters (*Homarus americanus*) caused by chitinovorous bacteria. *Journal of the Biological Board of Canada*, **3**, 358–62.

Hill, B.J. (1976). Properties of a virus isolated from the bivalve mollusc *Tellina tenuis* (da Costa). In *Wildlife Diseases*, L.A. Page (ed),pp. 445–52. Plenum Press: New York.

Hill, B.J. (1982). Infectious pancreatic necrosis virus and its virulence. In *Microbial Diseases of Fish*, R.J. Roberts (ed), pp. 91–114. Academic Press: London.

Hill, B.J. (1984). Importance of molluscan viral infections. In *Control of Virus Diseases*, E. Kurstak (ed), pp. 217–36. Marcel Dekker: New York.

Hitchner, E.R. and Snieszko, S.F. (1947). A study of a microorganism causing a bacterial disease of lobsters. *Journal of Bacteriology*, **54**, 48.

Horsley, R.W. (1973). The bacterial flora of the Atlantic salmon (*Salmo salar* L.) in relation to its environment. *Journal of Applied Bacteriology*, **36**, 377–86.

Hoshina,T. (1956). An epidemic disease affecting rainbow trout in Japan. *Journal of Tokyo University, Fisheries*, **42**, 15–6.

Jeffries, C.E. (1982).Three *Vibrio* strains pathogenic to larvae of *Crassostrea gigas* and *Ostrea edulis*. *Aquaculture*, **29**, 201–26.

Jensen, N.J. and Bloch, B. (1980). Adenovirus-like particles associated with epidermal hyperplasia in cod (*Gadus morhua*). *Nordische Veterinaer Medizine*, **32**, 173–5.

Jensen, N.J., Bloch, B. and Larsen, J.L. (1979).The ulcer syndrome in cod (*Gadus morhua*). III. A preliminary virological report. *Nordische Veterinaer Medizine*, **31**, 436–42.

Johnson, P.W., Sieburth, J. McN., Sastry, A., Arnold, C.R. and Doty, M.S. (1971). *Leucothrix mucor* infestation of benthic crustacea, fish eggs, and tropical algae. *Limnology and Oceanography*, **16**, 962–9.

Johnson, S.K. (1974). *Fusarium sp. in a laboratory-held pink shrimp*.Texas A & M University, Texas Agricultural Extension Service, Fish Disease Diagnostic Laboratory, Publication FDDL 1, 2pp.

Jollès, J. and Jollès, (1975).The lysozyme from *Asterias rubens*. *European Journal of Biochemistry*, **54**, 19–23.

Kang, J.W. (1981). Some seaweed diseases occurred at seaweed farms along the South-Eastern coast of Korea. *Bulletin of the Korean Fisheries Society*, **14**, 165–70.

Kashiwada, K., Kanazawa, A. and Teshima, S. (1971). Studies on the production of B vitamins by intestinal bacteria – VI. Production of folic acid by intestinal bacteria of carp. *Memoirs of the Faculty of Fisheries of Kagoshima University*, **20**, 185–9.

Kellogg, S., Steenbergen, J.F. and Schapiro, H.C. (1974). Isolation of *Pediococcus homari*, etiological agent of gaffkemia, from a California estuary. *Aquaculture*, **3**, 409–13.

Kerebel, B., Cabellec, M.T. le and Kerebel, L.M. (1979). Structure and ultrastructure of intra-vitam parasitic destruction of the external dental tissue in the fish, *Anarhichas lupus* L. *Archives of Oral Biology*, **24**, 147–53.

Kinne, O. (1980). *Diseases of Marine Animals*, vol. 1, *General Aspects, Protozoa to Gastropods*. Wiley: Chichester.

Kitao,T.,Aoki,T., Fukudome, M., Kawano, K.,Wada,Y. and Mizuno,Y. (1983). Serotyping of *Vibrio anguillarum* isolated from diseased freshwater fish in Japan. *Journal of Fish Diseases*, **6**, 175–81.

Laird, M. and Bullock,W.L. (1969). Marine fish hematozoa from New Brunswick and New England. *Journal of the Fisheries Research Board of Canada*, **26**, 1975–2002.

Langvad, F., Pedersen, O. and Engjom, K. (1985). A fungal disease caused by *Exophiala* sp. nova in farmed Atlantic salmon in western Norway. In *Fish and Shellfish Pathology*, A.E. Ellis (ed), pp. 323–8. Academic Press: London.

Larsen, J.L. (1982). *Vibrio anguillarum*: Prevalence in three carbohydrate loaded marine recipients and a control. *Zentralblatt für Bakteriologie und Hygiene*, **1**, **Abteilung Originale C3**, 519–30.

Laycock, R.A. (1974). The detrital food chain based on seaweeds. 1. Bacteria associated with the surface of *Laminaria* fronds. *Marine Biology*, **25**, 223–31.

Leibovitz, L. (1978). *A study of vibriosis at a Long Island Shellfish hatchery*. New York Sea Grant Reprint Series NYSG-RR-79-0-2.

Lightner, D.V. and Fontaine, C.T. (1973). A new fungus disease of the white shrimp *Penaeus setiferus*. *Journal of Invertebrate Pathology*, **22**, 94–9.

Liston, J. (1957). The occurrence and distribution of bacterial types on flatfish. *Journal of General Microbiology*, **16**, 205–16.

Lom, J. (1984). Diseases caused by protistans. In *Diseases of Marine Animals*, O. Kinne (ed), vol. 1, Part 1, pp. 114–68. Biologische Anstalt Helgoland: Hamburg.

Lovelace,T.E.,Tubiash, H. and Colwell, R.R. (1968). Quantitative and qualitative commensal bacterial flora of *Crassostrea virginica* in Chesapeake Bay. *Proceedings of the National Shellfish Association*, **58**, 82–7.

Lowe, J. (1874). *Fauna and Flora of Norfolk. Part IV. Fishes*. Transactions of the Norfolk and Norwich Naturalists Society, pp. 21–56.

MacDonald, N.L., Stark, J.R. and Austin, B. (1986). Bacterial microflora in the gastro-intestinal tract of Dover sole (*Solea solea* L.), with emphasis on the possible role of bacteria in the nutrition of the host. *FEMS Microbiology Letters*, **35**, 107–11.

Mattheis, T. (1964). Das Vorkommen von *Vibrio anguillarum* in Ostseefischen. *Zentralblatt für Fischerei*, **N.F. XII**, 259–63.

McAllister, P.E., Newman, M.W., Sauber, J.H. and Owens,W.J. (1983). Infectious pancreatic necrosis virus: isolation from Southern flounder, *Paralichthys lethostigma*, during an epizootic. *Bulletin of the European Association of Fish Pathologists*, **3**, 37–8.

McCain, B.B., Gronlund,W.D., Myers, M.S. andWellings, S.R. (1979).Tumours and microbial diseases of marine fishes in Alaskan waters. *Journal of Fish Diseases*,**2**, 111–30.

McGinnis, M.R. and Ajello, L. (1974). A new species of *Exophiala* isolated from channel catfish. *Mycologia*, **66**, 518–20.

McHenery, J.G., Birkbeck,T.H. andAllen, J.A. (1979).The occurrence of lysozyme in marine bivalves. *Comparative Biochemistry and Physiology*, **63B**, 25–8.

McVicar, A.H. (1982). Ichthyophonus infection of fish. In *Microbial Diseases of*

Fish, R.J. Roberts (ed), pp. 243–69. Academic Press: New York.

Meyers ,T.R. (1979). Reo-like virus isolated from juvenile American oysters (*Crassostrea virginica*). *Journal of General Virology*, **43**, 203–12.

Möller, H. (1974). *Ichthyosporidium hoferi* (Plehn et Muslon) (fungi) as parasite in the Baltic cod (*Gadus morhua*). *Kieler Meeresfischeri*, **30**, 37–41.

Möller, H. and Anders, K. (1986). *Diseases and Parasites of Marine Fishes.* Verlag Möller: Kiel.

Neish, G.A. and Hughes, G.C. (1980). Fungal diseases of fishes. In: *Diseases of Fishes*, S.F. Snieszko and H.R. Axelrod (eds), Book 6, T.F.H.: Neptune.

Nigrelli, R.F. and Vogel, H. (1963). Spontaneous tuberculosis in fishes and in other cold-blooded vertebrates with special reference to *Mycobacterium fortuitum* Cruz from fish and human lesions. *Zoologica, New York*, **48**, 130–43.

Nishibuchi, M. and Muroga, K. (1977). Pathogenic *Vibrio* isolated from cultured eels. 111. NaCl tolerance and flagellation. *Fish Pathology*, **12**, 87–92.

Nybelin, O. (1935). Untersuchungen über den bei Fischen Krankheitsegenden Spatpilz *Vibrio anguillarum*. *Meddelanden frau Statens Undersöknings -och Fasok-sanstalt für Sotvattenfisket Stockholm*, **8**, 1–62.

Otte, E. (1964). Eine Mykose bei einem Stachebrochen (*Trigon pastinaca* L.). *Wiener tierärztlichte Monatsschrifte*, **51**, 171–5.

Pacha, R.E. and Kiehn, E.D. (1969). Characterization and relatedness of marine vibrios pathogenic to fish: physiology, serology and epidemiology. *Journal of Bacteriology*, **100**, 1242–47.

Peyer, B. (1926). Über einen Fall von Caries an einem Rochengebiss. *Verhandlungen der Schweizerischen Naturforschenden*, **Gesellschaft 107**, 242.

Phillips, N.W. (1984). Role of different microbes and substrates as potential suppliers of specific essential nutrients to marine detritivores. *Bulletin of Marine Science*, **35**, 283–98.

Rabin, H. and Hughes, J.R. (1968). Studies on the host parasite relationship in gaffkemia. *Journal of Invertebrate Pathology*, **10**, 335–44.

Ransom, D.P. (1978). *Bacteriologic, immunologic and pathologic studies of* Vibrio *sp. pathogenic to salmonids*. Ph.D. Thesis, Oregon State University: Corvallis.

Reichenbach-Klinke, H.H. (1954). Untersuchungen über die bei Fischen durch Parasiten hervorgerufenen Zysten und deren Wirkung auf den Wirtskörper. 1. *Zentralblatt für Fischerei*, **3**, 565–636.

Reichenbach-Klinke, H.H. (1955). Beobacthungen über Meeresfisch-Tuberkulose. *Publ. Staz. zool. Napoli*, **26**, 55–62. *Publicaciones de Stazione Zoologica di Napoli*

Richards, R.H., Holliman, A. and Helgason, S. (1978). *Exophiala salmonis* infection in Atlantic salmon *Salmo salar* L. *Journal of Fish Diseases*, **1**, 357–68.

Ridley, S.C. and Slabyj, B.M. (1978). Microbiological evaluation of shrimp (*Pandalus borealis*) processing. *Journal of Food Protection*, **41**, 40–3.

Rieper, M. (1978). Bacteria as food for marine harpacticoid copepods. *Marine Biology*, **45**, 337–45.

Rosen, B. (1970). Shell disease of aquatic crustaceans. In *A Symposium on Diseases of Fishes and Shellfishes*, S.F. Snieszko (ed), Special Publication No. 5, pp. 409–15. American Fisheries Society: Washington, DC.

Rungger, D., Rostelli, M., Braendle, E. and Malsberger, R.G. (1971). A virus-like particle associated with lesions in the muscle of *Octopus vulgaris*. *Journal of Invertebrate Pathology*, **17**, 72–89.

Sakata,T., Nakaji, M. and Kakimoto, D. (1978). Microflora in the digestive tract of marine fish. 1. General characterization of the isolates from yellowtail. *Memoirs of the Faculty of Fisheries of Kagoshima University*, **27**, 65–71.

Sakata, T., Okabayashi, J. and Kakimoto, D. (1980). Variations in the intestinal microflora of *Tilapia* reared in fresh and seawater. *Bulletin of the Japanese Society of Scientific Fisheries*, **46**, 313–7.

Sano,T., Okamoto, N. and Nishimura,T. (1981). A new viral epizootic of *Anguilla japonica* Temminck and Schlegel. *Journal of Fish Diseases*, **4**, 127–39.

Schiewe, M.H. (1981). Taxonomic status of marine vibrios pathogenic for salmonid fish. *Developments in Biological Standardization*, **49**, 149–58.

Shiba,T. and Taga, N. (1978). Heterotrophic bacteria attached to seaweeds. *Journal of Experimental Marine Biology and Ecology*, **47**, 251–8.

Sieburth, J.McN., Brooks, R.D., Gessner, R.V., Thomas, C.D. and Tootle, J.L. (1974). Microbial colonization of marine plant surfaces as observed by scanning electron microscopy. In *Effect of the Ocean Environment on Microbial Activity*, R.R. Colwell and R.Y. Morita (eds), pp. 418–32. University Park Press: Baltimore.

Sieburth, J.McN. and Conover, J.T. (1965). *Sargassum* tannin, an antibiotic that retards fouling. *Nature (London)*, **108**, 52–3.

Skerman, V.B.D., McGowan, V. and Sneath, P.H.A. (1980). Approved lists of bacterial names. *International Journal of Systematic Bacteriology*, **30**, 225–420.

Sleeter, T.D., Boyle, P.J., Cundell, A.M. and Mitchell, R. (1978). Relationships between marine micro-organisms and the wood-boring isopod *Limnoria tripunctata*. *Marine Biology*, **45**, 329–36.

Smith, I.W. (1961). A disease of finnock due to *Vibrio anguillarum*. *Journal of General Microbiology*, **74**, 247–52.

Snieszko, S.F. and Taylor, C.C. (1947). A bacterial disease of the lobster (*Homarus americanus*). *Science (New York)*, **105**, 500.

Sochard, M.R., Wilson, D.F., Austin, B. and Colwell, R.R. (1979). Bacteria associated with the surface and gut of marine copepods. *Applied and Environmental Microbiology*, **37**, 750–9.

Stewart, J.E. (1972). The detection of *Gaffkya homari*, the bacterium pathogenic to lobsters (genus *Homarus*). *Fisheries Research Board of Canada*, **New Series 43**, 1–5.

Stewart, J.E. and Arie, B. (1974). Effectiveness of vancomycin against gaffkemia, the bacterial disease of lobsters (genus *Homarus*). *Journal of the Fisheries Research Board of Canada*, **31**, 1873–9.

Stewart, J.E. and Cornick, J.W. (1972). Effects of *Gaffkya homari* on glucose, total carbohydrates, and lactic acid of the hemolymph of the lobster (*Homarus americanus*). *Canadian Journal of Microbiology*, **18**, 1511–3.

Stewart, J.E., Cornick, J.W., Spears, D.I. and McLeese, D.W. (1966). Incidence of *Gaffkya homari* in natural lobster (*Homarus americanus*) populations of the Atlantic region of Canada. *Journal of the Fisheries Research Board of Canada*, **23**, 1325–30.

Tolmasky, M.E. and Crosa, J.H. (1984). Molecular cloning and expression of genetic determinants for the iron uptake system mediated by the *Vibrio anguillarum* plasmid pJM1. *Journal of Bacteriology*, **160**, 860–6.

Trust,T.J., Courtice, I.D., Khouri, A.G., Crosa, J.H. and Schiewe, M.H. (1981). Serum resistance and haemagglutination ability of marine vibrios pathogenic for fish. *Infection and Immunity*, **34**, 702–7.

Tsukidate, J. (1971). Microbiological studies of *Porphyra* plants. II. Bacteria isolated from *Porphyra leucostica* in culture. *Bulletin of the Japanese Society of Scientific Fisheries*, **37**, 376–9.

Tubiash, H.S., Chanley, P.E. and Liefson, E. (1965). Bacillary necrosis, a disease of larval and juvenile molluscs. *Journal of Bacteriology*, **90**, 1036–44.

Ugajin, M. (1979). Studies on the taxonomy of major microflora in the intestinal contents of salmonids. *Bulletin of the Japanese Society of Scientific Fisheries*, **45**, 721–31.

Underwood, B.O., Smale, C.J., Brown, F. and Hill, B.J. (1977). Relationship of a virus from *Tellina tenuis* to infectious pancreatic necrosis virus. *Journal of General Virology*, **36**, 93–109.

Unkles, S.E. (1977). Bacterial flora of the sea urchin *Echinus esculentus*. *Applied and Environmental Microbiology*, **34**, 347–50.

Vacelet, J. (1975). Etude en microscopie électronique de l'association entre bactéries et spongiaires de genre *Verongia* (Dictyoceratida), *Journal de Microscopie et Biologie Cellulares*, **23**, 271–88.

Vago, C. (1966). A virus disease in crustacea. *Nature (London)*, **209**, 1290.

Vanderzant, C. and Nickelson, R. (1970). Isolation of *Vibrio parahaemolyticus* from Gulf Coast shrimp. *Journal of Milk and Food Technology*, **33**, 161–2.

Van Duijn, C. (1981). Tuberculosis in fishes. *Journal of Small Animal Practice*, **22**, 391–411.

Von Betegh L. (1910). Weitere Beiträge zur experimentellen Tuberkulose der Meeresfischen. *Zentralblatt für Bakteriologie, Parasitenkunde, Infektionskrankheiten und Hygiene*, **Abteilung 1**, 53–4.

Wardlaw, A.C. and Unkles, S.E. (1978). Bactericidal activity of coelomic fluid from the sea urchin *Echinus esculentus*. *Journal of Invertebrate Pathology*, **32**, 25–34.

Wells, N.A. and Zobell, C.E. (1934). *Achromobacter ichthyodermis*, n. sp., the etiological agent of an infectious dermatitis of certain marine fishes. *Proceedings of the National Academy of Science USA*, **20**, 123–6.

Wilkinson, C.R. (1978a). Microbial associations in sponges. I. Ecology, physiology and microbial populations of coral reef sponges. *Marine Biology*, **49**, 161–7.

Wilkinson, C.R. (1978b). Microbial associations in sponges. II. Numerical analysis of sponge and water bacterial populations. *Marine Biology*, **49**, 169–76.

Wilkinson, C.R., Nowak, M., Austin, B. and Colwell, R.R. (1981). Specificity of bacterial symbionts in Mediterranean and Great Barrier reef sponges. *Microbial Ecology*, **7**, 13–21.

Winton, J.R., Lannan, C.N. and Fryer, J.L. (1983). Further characterization of a new reovirus of poikilothermic vertebrates. In: *Double-Stranded RNA Viruses*, R.W. Compass and D.H.L. Bishop, (eds), pp. 231–6. Elsevier: Amsterdam.

Zachary, A. and Colwell, R.R. (1979). Gut-associated microflora of *Limnoria tripunctata* in marine creosote-treated wood pilings. *Nature (London)*, **282**, 716–7.

Zobell, C.E. and Feltham, C.B. (1938). Bacteria as food for certain marine invertebrates. *Journal of Marine Research*, **1**, 312–25.

Zobell, C.E. and Landon, W.A. (1937). Bacterial nutrition of the California mussel. *Proceedings of the Society of Experimental Biology and Medicine*, **36**, 607–9.

Chapter 7

Arrhenius, G. (1952). *Sediment cores from the East Pacific. 1. Properties of the sediment and their distribution*. Report of the Swedish Deep-Sea Expedition 1947–1948.

Ballard, R.D. (1977). Notes on a major oceanographic find. *Oceanus*, **20**, 35–44.

Barber, R.T. (1968). Dissolved organic carbon from deep waters resists microbial oxidation. *Nature (London)*, **207**, 274–5.

Baross, J.A. and Deming, J.W. (1983). Growth of 'black smoker' bacteria at temperatures of at least 250 °C. *Nature (London)*, **303**, 423–6.

Baross, J.A., Lilley, M.D. and Gordon, L.I. (1982). Is the CH_4, H_2 and CO venting from submarine hydrothermal systems produced by thermophilic bacteria? *Nature (London)*, **298**, 366–8.

Baross, J.A., Tester, P.A. and Morita, R.Y. (1978). Incidence, microscopy and etiology of exoskeleton lesions in the Tanner crab, *Chionoecetes tanneri*. *Journal of the Fisheries Research Board of Canada*, **35**, 1141–9.

Carlucci, A.F. and Williams, P.M. (1978). Simulated *in situ* growth rates of pelagic marine bacteria. *Naturwissenschaften*, **65**, 541–2.

Cavanaugh, C.M. (1985). Symbioses and chemoautotrophic bacteria and marine invertebrates from hydrothermal vents and reducing sediments. *Biological Society of Washington Bulletin*, **6**, 373–88.

Certes, A. (1884a). Sur la culture, a l'abri des germes atmospheriques, des eaux et des sediments rapportes par les expeditions du Travailleur et du Talisman. *Comptes Rendus Academie Science, Paris*, **98**, 690–3.

Certes, A. (1884b). Note relative a l'action des hautes pressions sur la vitalite des micro-organismes d'eau douce et d'eau de mer. *Comptes Rendus Society Biologie*, **36**, 220–2.

Comita, P.B. and Gagosian, R.B. (1984). Membrane lipid from deep-sea hydrothermal vent methanogen: a new macrocyclic glycerol diether. *Science (Washington, DC)*, **222**, 1329–31.

Corliss, J.B., Dymond, J., Gordon, L.I., Edmond, J.M., von Herzen, R.P., Ballard, R.D., Green, K., Williams, D., Bainbridge, A., Crane, J. and van Andel, T.H. (1979). Submarine thermal springs on the Galapagos Rift. *Science (Washington, DC)*, **203**, 1073–83.

De Long, E.F. and Yayanos, A.A. (1986). Biochemical function and ecological significance of novel bacterial lipids in deep-sea procaryotes. *Applied and Environmental Microbiology*, **51**, 730–7.

Deming, J.W. (1981). *Ecology of barophilic deep-sea bacteria*. Ph.D. Thesis: University of Maryland.

Deming, J.W. (1985). Bacterial growth in deep-sea sediment trap and boxcore samples. *Marine Ecology–Progress Series*, **25**, 305–12.

Deming, J.W. and Baross, J.A. (1986). Solid medium for culturing black smoker bacteria at temperatures to 120 °C. *Applied and Environmental Microbiology*, **51**, 238–43.

Deming, J.W. and Colwell, R.R. (1985). Observations of barophilic microbial activity in samples of sediment and intercepted particulates from the Demerara Abyssal Plain. *Applied and Environmental Microbiology*, **50**, 1002–6.

Deming, J.W., Tabor, P.S. and Colwell, R.R. (1981). Barophilic growth of bacteria from intestinal tracts of deep-sea invertebrates. *Microbial Ecology*, **7**, 85–94.

Dietz, A.S. and Yayanos, A.A. (1978). Silica gel media for isolating and studying bacteria under hydrostatic pressure. *Applied and Environmental Microbiology*, **36**, 966–8.

Ehrlich, H. (1983). Manganese oxidizing bacteria from a hydrothermally active area on the Galapagos Rift. *Ecology Bulletin*, **35**, 357–66.

Fiala, G., Stetter, K.O., Jannasch, H.W., Langworthy, T.A. and Madon, J. (1986). *Staphylothermus marinus* sp. nov. represents a novel genus of extremely thermophilic submarine heterotrophic Archaebacteria growing up to 98 °C. *Systematic and Applied Microbiology*, **8**, 106–13.

Fischer, B. (1894). Die Bakterien des Meeres nach den Untersuchungen der Plankton. Expedition unter gleichzeitiger Berucksichtigung einiger alterer und neuerer Untersuchungen. *Zentralblatt für Bakteriologie*, **15**, 657–66.

Fleminger, A. (1983). Description and phylogeny of *Isaacsicalanus paucisetus*, gen. and sp. n. (Copepoda, Calanoida, Spinocalanidae) from an East Pacific hydrothermal vent site (21 ° N). *Proceedings of the Biological Society of Washington*, **96**, 605–22.

Honjo, S. (1978). Sedimentation of materials in the Sargasso Sea at a 5367 m station. *Journal of Marine Research*, **36**, 469–92.

Isaacs, J.D. (1969). The nature of oceanic life. *Scientific American*, **221**, 146–62.

Jannasch, H.W. (1984). Microbes in the oceanic environment. In *The Microbe 1984*, D.P. Kelly and N.G. Carr (eds), part II, pp. 97–122. Cambridge University Press: Cambridge.

Jannasch, H.W. (1985). The chemosynthetic support of life and the microbial diversity at deep-sea hydrothermal vents. *Proceedings of the Royal Society of London*, **B225**, 277–97.

Jannasch, H.W., Eimhjellen, K., Wirsen, C.O. and Farmanfarmaian, A. (1971). Microbial degradation of organic matter in the deep sea. *Science (Washington, DC)*, **171**, 672–5.

Jannasch, H.W. and Mottl, M.J. (1985). Geomicrobiology of deep sea hydrothermal vents. *Science (Washington, DC)*, **229**, 717–25.

Jannasch, H.W. and Nelson, D.C. (1984). Recent progress in the microbiology of hydrothermal vents. In *Current Perspectives in Microbial Ecology*, C.A. Reddy and M.J. Klug (eds), pp. 170–6. American Society for Microbiology: Washington, DC.

Jannasch, H.W. and Taylor, C.D. (1984). Deep-sea microbiology. *Annual Review of Microbiology*, **38**, 487–514.

Jannasch, H.W. and Wirsen, C.O. (1973). Deep-sea microorganisms: *in situ* response to nutrient enrichment. *Science (Washington, DC)*, **180**, 641–3.

Jannasch, H.W. and Wirsen, C.O. (1979). Chemosynthetic primary production at East Pacific sea floor spreading centers. *Bioscience*, **29**, 592–8.

Jannasch, H.W. and Wirsen, C.O. (1981). Morphological survey of microbial mats near deep-sea thermal vents. *Applied and Environmental Microbiology*, **41**, 528–38.

Jannasch, H.W. and Wirsen, C.O. (1982). Microbial activities in undercompressed and decompressed deep-seawater samples. *Applied and Environmental Microbiology*, **43**, 1116–24.

Jannasch, H.W., Wirsen, C.O., Cuhel, R.L. and Taylor, C.D. (1980). An approach for *in situ* studies of deep-sea amphipods and their microbial gut flora. *Deep-Sea Research*, **27**, 867–72.

Jannasch, H.W., Wirsen, C.O., Nelson, D.C. and Robertson, L.A. (1985). *Thiomicrospira crunogena* sp. nov., a colourless sulphur oxidizing bacterium from a deep sea hydrothermal vent. *International Journal of Systematic Bacteriology*, **35**, 422–4.

Jannasch, H.W., Wirsen, C.O. and Taylor, C.D. (1976). Undecompressed microbial populations from the deep sea. *Applied and Environmental Microbiology*, **32**, 360–7.

Jones, W.J., Leigh, J.A., Mayer, F., Woese, C.R. and Rolfe, R.S. (1983). *Methanococcus jannaschii* sp. nov., an extremely thermophilic methanogen from a submarine hydrothermal vent. *Archives of Microbiology*, **136**, 254–61.

Karl, D.M. (1980). Cellular nucleotide measurements and applications in microbial ecology. *Microbiological Reviews*, **44**, 739–96.

Karl, D.M. (1987). Bacterial production at deep-sea hydrothermal vents and cold seeps: evidence for chemosynthetic primary production. In *Ecology of Microbial Communities*, M. Fletcher, T.R.G. Gray and J.G. Jones (eds), pp. 319–60. Cambridge University Press: Cambridge.

Karl, D.M., Wirsen, C.O. and Jannasch, H.W. (1980). Deep-sea primary production at the Galapagos hydrothermal vent. *Science (Washington, DC)*, **207**, 1345–7.

Kuenen, J.G. and Veldkamp, H. (1972). *Thiomicrospira pelophila*, gen. n., sp. n., a new obligately chemolithotrophic colourless sulfur bacterium. *Antonie van Leeuwenhoek*, **38**, 241–56.

Lonsdale, P. (1977). Clustering of suspension-feeding macrobenthos near abyssal hydrothermal vents at oceanic spreading centers. *Deep-Sea Research*, **24**, 857–63.

Menzel, D.W. and Ryther, J.H. (1970). Distribution and cycling of organic matter in the oceans. In *Organic Matter in Natural Waters*, D.W. Hood (ed), pp. 31–54. Institute of Marine Science Publications, College, A.K..

Morita, R.Y. (1979). The role of microbes in the bioenergetics of the deep-sea. *Sarsia*, **64**, 9–12.

Morita, R.Y. (1980). Microbial life in the deep sea. *Canadian Journal of Microbiology*, **26**, 1375–85.

Paul, K.L. and Morita, R.Y. (1971). The effects of hydrostatic pressure and temperature on the uptake and respiration of amino acids by a facultatively psychrophilic marine bacteria. *Journal of Bacteriology*, **108**, 835–43.

Quigley, M.M. and Colwell, R.R. (1968). Properties of bacteria isolated from deep-sea sediments. *Journal of Bacteriology*, **95**, 211–20.

Ruby, E.G. and Jannasch, H.W. (1982). Physiological characteristics of *Thiomicrospira* sp. L-12 isolated from deep sea hydrothermal vents. *Journal of Bacteriology*, **149**, 161–5.

Ruby, E.G., Wirsen, C.O. and Jannasch, H.W. (1981). Chemolithotrophic sulfur-oxidizing bacteria from the Galapagos Rift hydrothermal vents. *Applied and Environmental Microbiology*, **42**, 317–24.

Schwartz, J.R., Walker, J.D. and Colwell, R.R. (1974). Growth of deep-sea bacteria on hydrocarbons at ambient and *in situ* pressure. *Developments in Industrial Microbiology*, **15**, 239–49.

Seki, H., Wada, E., Koike, I. and Hattori, A. (1974). Evidence of high organotrophic potentiality of bacteria in the deep ocean. *Marine Biology,* **26**, 1–4.

Silver, M.W., Shanks, A.L. and Trent, J.D. (1978). Marine snow: microplankton habitat and source of small-scale patchiness in pelagic populations. *Science (Washington, DC),* **201**, 371–3.

Stahl, D.A., Lane, D.J., Olsen, G.L. and Pace, N.R. (1984). Analysis of hydrothermal vent-associated symbionts in ribosomal RNA sequences. *Science (Washington, DC),* **224**, 409–11.

Suess, E. (1980). Particulate organic carbon flux in the oceans–surface productivity and oxygen utilization. *Nature (London),* **288**, 260–3.

Tabor, P.S., Deming, J.W., Ohwada, K. and Colwell, R.R. (1982). Activity and growth of microbial populations in pressurized deep-sea sediments and animal gut samples. *Applied and Environmental Microbiology,* **44**, 413–22.

Tabor, P.S., Deming, J.W., Ohwada, K., Davis, H., Waxman, M. and Colwell, R.R. (1981a). A pressure-retaining deep ocean sampler and transfer system for measurement of microbial activity in the deep sea. *Microbial Ecology,* **7**, 51–65.

Tabor, P.S., Ohwada, K. and Colwell, R.R. (1981b). Filterable marine bacteria found in the deep sea: distribution, taxonomy, and response to starvation. *Microbial Ecology,* **7**, 67–83.

Turner, J.T. (1979). Microbial attachment to copepod fecal pellets and its possible ecological significance. *Transactions of the American Microscopic Society,* **98**, 131–5.

Tuttle, J.H., Wirsen, C.O. and Jannasch, H.W. (1983). Microbial activities in the emitted hydrothermal waters of the Galapagos Rift vents. *Marine Biology,* **73**, 293–9.

Vinogradov, M.E. and Tseitlin, V.B. (1983). Deep sea pelagic domain (aspects of the bioenergetics). In *Deep-Sea Biology,* G.T. Rowe (ed), pp. 123–65. John Wiley and Sons: New York.

Wada, E., Koike, I. and Hattori, A. (1975). Nitrate metabolism in abyssal waters. *Marine Biology,* **29**, 119–24.

Weiner, R.M., Devine, R.A, Powell, D.M., Dagasan, L. and Moore, R.L. (1985). *Hyphomonas oceanitis* spec. nov., *H. hirschiana* spec. nov. and *H. jannaschiana* spec. nov. *International Journal of Systematic Bacteriology,* **35**, 237–43.

Wirsen, C.O. and Jannasch, H.W. (1983). *In situ* studies on deep sea amphipods and their intestinal microflora. *Marine Biology,* **78**, 69–73.

Yayanos, A.A., Dietz, A.S. and Van Boxtel, R. (1979). Isolation of a deep-sea barophilic bacterium and some of its growth characteristics. *Science (Washington, DC),* **205**, 808–10.

Yayanos, A.A., Dietz, A.S. and Van Boxtel, R. (1981). Obligately barophilic bacterium from the Marianas Trench. *Proceedings of the National Academy of Science USA,* **78**, 5212–5.

Yayanos, A.A, Dietz, A.S. and Van Boxtel, R. (1982). Dependence of reproduction rate on pressure as a hallmark of deep-sea bacteria. *Applied and Environmental Microbiology,* **44**, 1356–61.

Zobell, C.E. and Johnson, F.H. (1949). The influence of hydrostatic pressure on the growth and viability of terrestrial and marine bacteria. *Journal of Bacteriology,* **57**, 179–89.

Zobell, C.E. and Morita, R.Y. (1959). Deep-sea bacteria. *Galathea Report,* **1**, 139–54.

Chapter 8

Alexander, B. and Austin, B. (1986). Bacterial microflora associated with a commercial fish smoker. *FEMS Microbiology Letters*, **34**, 309–12.

Atlas, R.M. (1981). Microbial degradation of petroleum hydrocarbons: an environmental perspective. *Microbiological Reviews*, **45**, 180–209.

Beastall, S. (1977). Treatment and natural fate of oil spillages. In *Treatment of Industrial Effluents*, A.G. Calley, C.F. Forster and D.A. Stafford (eds), pp. 328–35. Hodder and Stoughton: London.

Berk, S.G. and Colwell, R.R. (1981). Transfer of mercury through a marine microbial food web. *Journal of Experimental Marine Biology and Ecology*, **52**, 157–72.

Bissett, K.A. (1948). Natural antibodies in the blood serum of freshwater fish. *Journal of Hygiene, Cambridge*, **46**, 267–8.

Bonar, D.B., Weiner, R.M. and Colwell, R.R. (1986). Microbial–invertebrate interactions and potential for biotechnology. *Microbial Ecology*, **12**, 101–10.

Chen, H.-C. and Chai,T.-J. (1982). Microflora of drainage from ice in fishing vessel fish holds. *Applied and Environmental Microbiology*, **43**, 1360–5.

Colwell, R.R., Belas, M.R., Zachary, A., Austin, B. and Allen, D.A. (1980). Attachment of micro-organisms to surfaces in the aquatic environment. *Developments in Industrial Microbiology*, **21**, 169–78.

Colwell, R.R., Berk, S.G., Sayler, G.S., Nelson, J.D. and Esser, J.M. (1975). Mobilization of mercury by aquatic microorganisms. In *International Conference on Heavy Metals in the Environment*, pp. 831–44. Toronto, Ontario, Canada.

Cowen, J.P. and Silver, M.W. (1984). The association of iron and manganese with bacteria on marine macroparticulate material. *Science (Washington, DC)*, **224**, 1340–2.

Crisp, D.J. (1974). Factors influencing the settlement of marine invertebrate larvae. In *Chemoreception in Marine Organisms*, P.T. Grant and A.M. Macke (eds), pp. 177–265. Academic Press: New York.

Dicks, B. (1982). Monitoring the biological effects of North Sea platforms. *Marine Pollution Bulletin*, **13**, 221–7.

Dicks, B. and Hartley, J.P. (1982). The effects of repeated small oil spillages and chronic discharges. *Philosophical Transactions of the Royal Society, London*, **B 297**, 285–307.

El-Sahn, M.A., El-Banna, A.A. and El-Tabey Shehata, A.M. (1982). Occurrence of *Vibrio parahaemolyticus* in selected marine invertebrates, sediment, and seawater around Alexandria, Egypt. *Canadian Journal of Microbiology*, **28**, 1261–4.

Gibson, W.L. and Brown, L.R. (1975). The metabolism of parathion by *Pseudomonas aeruginosa*. *Developments in Industrial Microbiology*, **16**, 77–87.

Horowitz, A. and Atlas, R.M. (1980). Microbial seeding to enhance petroleum hydrocarbon biodegradation in aquatic arctic ecosystems. In *Biodeterioration: Proceedings of the Fourth International Symposium, Berlin*, T.A. Oxley, D. Allsopp and G. Becker (eds), pp. 15–20. Pitman: London.

Hughes, D.E. and McKenzie, P. (1975). The microbial degradation of oil at sea. *Proceedings of the Royal Society, London*, **B 189**, 375–90.

Johnston, C.S. (1980). Sources of hydrocarbons in the marine environment. In *Oily*

Water Discharges, C.S. Johnston and R.J. Morris (eds), pp. 41–62. Applied Science: London.

Kirchman, D.L., Graham, D.S., Reish, D. and Mitchell, R. (1982a). Bacteria induce settlement and metamorphosis of *Janua (Dexiospira) brasiliensis* (Grube). *Journal of Experimental Marine Biology and Ecology*, **56**, 153–63.

Kirchman, D.L., Graham, D.S., Reish, D. and Mitchell, R. (1982b). Lectins may mediate in the settlement and metamorphosis of *Janua (Dexiospira) brasiliensis* (Grube). *Marine Biology Letters*, **3**, 131–42.

Mori, K., Shinano, H. and Akiba, M. (1977). Studies on the micro-organisms in salted and ripened squid meat product ("Ika-Shiokara") – 1.Yeasts in ripening process of "Ika-Shiokara". *Bulletin of the Japanese Society of Scientific Fisheries*, **43**, 1425–32.

Müller, G. (1977). Befunde an nicht-agglutinierenden, cholera-ähnlichen Vibrionen (NAG's) in Abwasser, Flusswasser und Meerwasser. *Zentralblatt für Bakteriologie und Hygiene*, **1 Abteilung, Originale B, 165**, 487–97.

Müller,W. and Buchal, G. (1973). Metamorphose-Induktion bei Planulalarven. II. Induktion durch monovalente Kationen: die Bedeutung des Gibbs-Donnan-Verhaltnisses und der Na^+/K^+-ATPase. *Wilhelm Roux's Archiv*, **173**, 122–35.

Nealson, K.H. and Tebo, B. (1980). Structural features of manganese precipitating bacteria. *Origins of Life*, **10**, 117–26.

Neumann, R. (1979). Bacterial induction of settlement and metamorphosis in the planula larvae of *Cassiopea andromeda* (Cnidaria: Scyphozoa, Rhizostomea). *Marine Ecology – Progress Series*, **1**, 21–8.

Okuzumi, M., Yamanaka, H. and Kubozuka, T. (1984). Occurrence of various histamine-forming bacteria on/in fresh fishes. *Bulletin of the Japanese Society of Scientific Fisheries*, **50**, 161–7.

Pritchard, P.H. and Starr,T.J. (1973). Microbial degradation of oil and hydrocarbons in continuous culture. In *The Microbial Degradation of Oil Pollutants*, D.G. Ahearn and S.P. Meyers (eds), pp. 39–45. Louisiana State University, Center for Wetland Resources: Baton Rouge, LSU-SG-73-01.

Ramibeloarisoa, E., Rontani, J.F., Giusti, G., Duvnjak, Z. and Bertrand, J.C. (1984). Degradation of crude oil by a mixed population of bacteria isolated from sea-surface foams. *Marine Biology*, **83**, 69–81.

Raymond, D. (1974). *Metabolism of methylnaphthalenes and other related aromatic hydrocarbons by marine bacteria*. Ph.D. Dissertation, Rutgers University.

Roper, M.M. and Marshall, K.C. (1978). Biological control agents of sewage bacteria in marine habitats. *Australian Journal of Marine and Freshwater Research*, **29**, 335–43.

Rosenberg, E. and Gutnick, D.L. (1981). The hydrocarbon-oxidizing bacteria. In *The Prokaryotes A Handbook on Habitats, Isolation and Identification of Bacteria*, M.P. Starr, H. Stolp, H.G.Trüper, A. Balows and H.G. Schlegel (eds), volume I, pp. 903–12. Springer Verlag: New York.

Scheltema, R.S. (1974). Biological interactions determining larval settlement of marine invertebrates. *Thallasia Jugo*, **10**, 263–96.

Shewan, J.M. (1971). The microbiology of fish and fishery products – a progress report. *Journal of Applied Bacteriology*, **34**, 299–315.

Soli, G. (1973). Marine hydrocarbonoclastic bacteria: types and range of oil degradation. In *The Microbial Degradation of Oil Pollutants*, D.G.Ahearn and

S.P. Meyers (eds), pp. 141–6. Louisiana State University, Center for Wetland Resources: Baton Rouge, LSU-SG-73-01.

Soli, G. and Bens, E.M. (1972). Bacteria which attack petroleum hydrocarbons in a saline environment. *Biotechnology and Bioengineering*, **14**, 319–30.

Thi-Son, N. and Fleet, G.H. (1980). Behavior of pathogenic bacteria in the oyster, *Crassostrea commercialis*, during depuration, re-laying and storage. *Applied and Environmental Microbiology* **40**, 994–1002.

Van den Broek, M.J.M., Mossel, D.A.A. and Eggenkamp. A.E. (1979). Occurrence of *Vibrio parahaemolyticus* in Dutch mussels. *Applied and Environmental Microbiology*, **37**, 438–42.

Van Spreekens, K.J.A. (1977). Characterization of some fish and shrimp spoiling bacteria. *Antonie van Leeuwenhoek*, **43**, 283–303.

Vance, I., Stanley, S.O. and Brown, C.M. (1979). A microscopical investigation of the bacterial degradation of wood pulp in a simulated marine environment. *Journal of General Microbiology*, **114**, 69–74.

Voronin, A.M., Kochetkov, V.V., Starovoytov, I.I. and Skryabin, G.K. (1977). pBs2 and pBS3 plasmids controlling oxidation of naphthalene in bacteria of the genus *Pseudomonas*. *Moscow Doklady Akademii Nauk*, **237**, 1205–8.

Walker, J.D., Colwell, R.R. and Petrakis, L. (1976). Biodegradation rates of components of petroleum. *Canadian Journal of Microbiology*, **22**, 1209–13.

Weiner, R.M. and Colwell, R.R. (1982). *Induction of settlement and metamorphosis in* Crassostrea virginica *by a melanin synthesizing bacterium*. Technical Report, Maryland Sea Grant, University of Maryland, UMSG-TS-82-05.

Whittle, K.J., Hardy, R., Mackie, P.R. and McGill, A.S. (1982). A quantitative assessment of the sources and fate of petroleum compounds in the marine environment. *Philosophical Transactions of the Royal Society*, **B**, **297**, 193–218.

Wood, E.J.F. (1967). *Microbiology of Oceans and Estuaries*. Amsterdam: Elsevier.

Zobell, C.E. (1969). Microbial modification of crude oil in the sea. In *API/FWPCA Conference on Prevention and Control of Oil Spills*, pp. 317–26. American Petroleum Institute, Washington.

Zobell, C.E. (1973). Microbial degradation of oil: present status, problems and perspectives. In: *The Microbial Degradation of Oil Pollutants*, D.G. Ahearn and S.P. Meyers (eds), pp. 3–16. Louisiana State University, Center for Wetland Resources: Baton Rouge, LSU-SG-73-01.

Zobell, C.E. and Allen, E.C. (1935). The significance of marine bacteria in the fouling of submerged surfaces. *Journal of Bacteriology*, **29**, 239–51.

Zobell, C.E. and Prokup, J.F. (1966). Microbial oxidation of mineral oils in Barataria Bay bottom deposits. *Zeitschrift für Allgemeine Mikrobiologie*, **6**, 134–62.

Chapter 9

Anderson, R.J., Wolfe, M.S. and Faulkner, D.J. (1974). Autotoxic antibiotic production by a marine *Chromobacterium*. *Marine Biology*, **27**, 281–5.

Baarn, R.B., Gandhi, N.M. and Freitas, Y.M. (1966). Antibiotic activity of marine microorganisms. *Helgoländer wissenschaftliche Meeresuntersuchungen*, **13**, 181–7.

Baumann, P., Gauthier, M.J. and Baumann, L. (1984). Genus *Alteromonas*

Baumann, Baumann, Mandel and Allen 1972, 418[AL]. In: *Bergeys Manual of Systematic Bacteriology*, N.E. Krieg (ed), vol. 1, pp. 343–52. Williams & Wilkins: Baltimore.

Bauwens, M. and De Ley, J. (1981). Improvements in the taxonomy of *Flavobacterium* by DNA:rRNA hybridizations. In *The Flavobacterium–Cytophaga Group*, H. Reichenbach and O.B Weeks (eds), pp. 27–31. Verlag Chemie: Weinheim.

Bonar, D.B., Weiner, R.M. and Colwell, R.R. (1986). Microbial–invertebrate interactions and potential for biotechnology. *Microbial Ecology*, **12**, 101–10.

Buck, J.D., Meyers, S.P. and Kamp, K.M. (1962). Marine bacteria with antiyeast activity. *Science (New York)*, **138**, 1339–40.

Burkholder, P.R., Pfister, R.M. and Leitz, F.H. (1966). Production of a pyrrole antibiotic by a marine bacterium. *Applied Microbiology*, **14**, 649–53.

De Giaxa (1889). Über das Verhalten einiger pathogener Mikroorganismen in Meerwasser. *Zentralblatt für Hygiene und Infektionkrankheiten*, **6**, 162–225.

Deming, J.W. (1986). The biotechnological future for newly described, extremely thermophilic bacteria. *Microbial Ecology*, **12**, 111–9.

Doggett, R.G. (1968). New anti-*Pseudomonas* agent isolated from a marine vibrio. *Journal of Bacteriology*, **95**, 1972–3.

Gauthier, M.J. and Flateau, G.N. (1976). Antibacterial activity of marine violet-pigmented *Alteromonas* with special reference to the production of brominated compounds. *Canadian Journal of Microbiology*, **22**, 1612–9.

Imada, C., Maeda, M., Hara, S., Taga, N. and Simidu, U. (1986). Purification and characterization of subtilism inhibitors 'Marinostatin' produced by marine *Alteromonas* sp. *Journal of Applied Bacteriology*, **60**, 469–76.

Katzenelson, E. (1978). Survival of viruses. In *Indicators of Viruses in Food and Water*, G. Berg (ed), pp. 39–50. University of Michigan Press: Ann Arbor, Michigan.

Kirchman, D.L., Graham, D.S., Reish, D. and Mitchell, R. (1982a). Bacteria induce settlement and metamorphosis of *Janua (Dexiospira) brasiliensis* Grube. *Journal of Experimental Marine Biology and Ecology*, **56**, 153–63.

Kirchman, D.L., Graham, D.S., Reish, D. and Mitchell, R. (1982b). Lectins may mediate in the settlement and metamorphosis of *Janua (Dexiospira) brasiliensis* Grube. *Marine Biology Letters*, **3**, 131–42.

Krassil'nikova, E.N. (1961). Antibiotic properties of microorganisms isolated from various depths of world oceans. *Mikrobiologiya*, **30**, 545–50.

Lemos, M.L., Toranzo, A.E. and Barja, J.L. (1985). Antibiotic activity of epiphytic bacteria isolated from intertidal seaweeds. *Microbial Ecology*, **11**, 149–63.

Lovell, F.M. (1966). The structure of a bromine-rich antibiotic. *Journal of the American Chemical Society*, **88**, 4510–1.

Magnussen, S., Gundersen, K., Brandenberg, A. and Lycke, E. (1967). Marine bacteria and their relationships to the virus inactivation capacity of seawater. *Acta Pathologia et Microbiologia, Scandinavia*, **71**, 274–80.

Neumann, R. (1979). Bacterial induction of settlement and metamorphosis in the planula larva of *Cassipea andromeda* (Cnidaria: Scyphozoa, Rhizostomaea). *Marine Ecology – Progress Series*, **1**, 21–8.

Okami, Y. (1986). Marine microorganisms as a source of bioactive agents. *Microbial Ecology*, **12**, 65–78.

Patel, R.N. and Hou, C.T. (1983). Enzymatic transformation of hydrocarbons by methanotrophic organisms. *Developments in Industrial Microbiology*, **23**, 187–205.

Phaff, H.J. (1986). Ecology of yeasts with actual and potential value in biotechnology. *Microbial Ecology*, **12**, 31–42.

Rosenfeld, W.D. and Zobell, C.E. (1947). Antibiotic production by marine microorganisms. *Journal of Bacteriology*, **54**, 393–8.

Skerman, V.B.D., McGowan, V. and Sneath, P.H.A. (1980). Approved lists of bacterial names. *International Journal of Systematic Bacteriology*, **30**, 225–420.

Staley, J.T. and Stanley, P.M. (1986). Potential commercial applications in aquatic microbiology. *Microbial Ecology*, **12**, 79–100.

Toranzo, A.E., Barja, J.L. and Hetrick, F.M. (1982). Antiviral activity of antibiotic-producing marine bacteria. *Canadian Journal of Microbiology*, **28**, 231–8.

Wratten, S.J., Wolfe, M.S., Anderson, R.J. and Faulkner, D.J. (1977). Antibiotic metabolites from a marine pseudomonad. *Antimicrobial Agents and Chemotherapy*, **11**, 411–4.

Zosin, Z., Gutnick, D. and Rosenberg, E. (1982). Properties of hydrocarbon-in-water emulsions stabilized by *Acinetobacter* RAG-1 emulsan. *Biotechnology and Bioengineering*, **24**, 281–92.

Zosin, Z., Gutnick, D. and Rosenberg, E. (1983). Uranium binding by emulsan and emulsanols. *Biotechnology and Bioengineering*, **25**, 1725–35.

Index

abyssal plains, 2
abyssopelagic, 11
acetogenesis, 97
acridine orange, 19
aerobic heterotrophs, numbers, 32
agar deeps, 27
akinetes, 51
algal pathogens, 124
algal productivity, 101
allergies, 156
Alvin, 13, 16, 136, 147
aminoglycosides, 167
ammonia oxidisers, 60
ammonification, 90
anoxygenic photosynthesis, 47
Antarctic Bottom Water, 7
Antarctic Convergence Water, 80
Antarctic Convergence Zone, 44
Antarctic Intermediate Water, 7
antibiotics, 157
    extraction methods, 159, 160, 162, 168
antimalarial compound, 167
antitumour compounds, 171
antiviral compounds, 170
aphotic, 6
Aplasmomycin, 167–9
artificial seawater, 24
assimilation (nutrients), 90
attachment mechanisms, 88, 89
attachment structures, 89
attractants (chemotaxis), 88
autochthonous, 80

bacterial counts, 32
bacterial pathogens, 115, 118, 123
bacterial productivity, 101
bacteriophage, 132
bacterioplankton, definition, 9
baeocytes, 51
barophiles, definition, 135
barotolerant bacteria, 133, 146
bathypelagic, 11
benefits, 149
benthic habitat, 11
benthos, definition, 11

biodegradation, 147
    pollutants, 149
biodeterioration, 153
biofouling community, 20, 88, 153
biogenous sediments, 8
biogeochemical processes, 90
    carbon cycle, 90, 96
    iron cycle, 99
    manganese cycle, 99
    nitrogen cycle, 90
    phosphorus cycle, 90, 99
    sulphur cycle, 90, 93
biomass, 19
biomass determination, 42
    adenosine triphosphate, 21, 42, 140
    bacteriochlorophyll, 21, 22
    chitin, 21
    chlorophyll, 21, 22
    DNA, 21
    lipopolysaccharide, 21, 22, 42
    muramic acid, 21
    phospholipids, 21, 22
biomass, microbial, 84
biotechnology, 157–74
black smoker bacteria, 137
black smokers, 135
'bottom landers', 138
brominated antibiotics, 163
bromopyrroles, 159–61

calcareous sediment, 8
carbon assimilation, 100
carbon cycle, 96
carbon dioxide fixation, 138
carbon dioxide incorporation, 147
carbon production, 100
chemolithotrophs, 60, 94
chemosynthesis, 148
chemotaxis, 88, 103
^{14}C-labelled substrates, 82, 85, 137, 146
colonisation,
    deep sea, 138
    surfaces, 88, 105, 107
continental shelf, 2, 8
continental slope, 2

corers, 13
coryneforms, 65
culture techniques, 22
    eukaryotes, 29
    micro-algae, 29
    prokaryotes, 22
cycling of elements, 90

Daro model, 86
deep sea, definition, 133
deep-sea organisms, role, 146
deep zone, 6
degradation, complex molecules, 87
    amylase, 87
    DNA, 87
    lipids, 87
    mannan, 87
    pollutants, 87
    protease, 87
detritivores, 147
direct counts, 32, 79
direct viable counts, 32, 41, 79
disease, definition, 118
diseases,
    vertebrates, 115
    invertebrates, 123
dissimilation, 90
dormancy, 79, 81
    constitutive, definition, 81
    exogenous, definition, 81
dormant cells, 40
dredge, 13, 18

ecological niche, 107
electron acceptors, 98
electrostatic repulsion, 89
Emulsan, 173
endobiotic habitat, definition, 9
energy production, 174
enzootic, definition, 118
enzymes, biotechnology, 172
epibiotic
    associations, 129
    habitat, definition, 9
epifluorescence microscopy, 19, 34, 40–1, 82,
    139–40
epiphytes, 104
epizootic, definition, 118
exophthalmia, 120
exopolymer, 152

femtoplankton, definition, 9
fermented foods, 153
ferromanganese deposits, 144
fish, microflora, 105
fish spoilage, 156
fluorescent antibody technique, 19
fluorochrome, 19

food chains
    decomposer, 84
    grazing, 84
food poisoning, 156
food webs, 83, 96, 152, 154
frustule, definition, 66
fungal pathogens, 116, 120, 129
future developments, 176–7

gaffkemia, 127
gastro-intestinal tract, microflora, 106–10,
    113, 140
Gelrite, 137
geothermal energy, 148
gill hyperplasia, 121
glucan degrading enzyme, 173
glutamine synthase, 92
glutamine synthetase, 92
grab sampler, 13, 17
granuloma, 120–1
grazers, 96
grazing, 34, 84
    rates, 85

heavy metals, mobilisation, 154
heterocysts, 51, 91
hormogonia, 51
histamine forming bacteria, 156
hot smoker chimney, 144
hot vents, 135
human pathogens, reservoir, 154
hydrocarbonoclastic organisms, 149
hydrocarbon degradation, 149
hydrogenous sediment, 8
hydrothermal fluid, 135
hydrothermal vent, 134–5
    microflora, 140
hyperplasia, epidermal, 122
    gill, 121

Ika–Shiokara, 153
immunogenicity, 119
iron-binding proteins, 119
iron
    cycle, 99
    sequestering mechanism, 119
isolation, mycobacteria, 120
Istamycin, 167, 171

lactose broth, 28
Limulus amoebocyte lysate assay, 22
lithogenous sediment, 8
Lotka–Volterra equations, 86
luciferin–luciferase reaction, 21
luminous bacteria, 38
lysis, fungi, 83
lysozyme, 114

macro-organism, 103
macroplankton, definition, 9
magnetite, 89
magnetotactic bacteria, 89
maintenance energy, 81
malefits, 153
manganese
  cycle, 99
  nodules, 8, 152
  oxidisers, 144
Marinactan, 172
marine environment, 1
marine
  particles, 98
  salts solution, 24
marine snow, 134
  definition, 85
Marinostatin, 167
mathematical modelling, 85
  Daro model, 86
mean generation time, 146
media recipes, 23–30
media
  Aerococcus viridans, 129
  antibiotic producers, 159, 161
  Chromobacterium spp., 165
  ciliates, 29
  fungi, 29
  methanogens, 27
  micro-algae, 29
  Photobacterium, 27
  phototrophs, 26
  Pseudomonas bromoutilis, 159
  Streptomyces tenjimariensis, 170
  sulphate reducers, 26
megaplankton, definition, 9
mesohyl, 113
mesopelagic, 11
mesophile, 7
mesoplankton, definition, 9
messenger weight, 13
methane precursor, 98
methanogenesis, 97–8
methanogens, 39, 62, 95
methylation, 96
methylotroph, 99
microbial mat, 140
microflora
  deep-sea, 138
  diseased invertebrates, 118
  diseased vertebrates, 115
  healthy invertebrates, 111
  healthy vertebrates, 105
  plants, 104
microplankton, definition, 9
microscopy, 15
mineralisation, 98
minibacteria, 81

minicells, 40–1
mobilisation of metals, 154
most probable numbers technique, 28
mud, 8
mutualism, definition, 9
mycobacteriosis, 120

nalidixic acid technique, 32, 41, 79, 82
nanoplankton, 9, 44
  heterotrophic, 84
Nansen bottle, 13, 15
nekton, definition, 9
neoplastic condition, 122
nets, 13–14
neuston
  definition, 9
  sampler, 14
Niskin sampler, 13, 16
nitrification, 92
nitrite oxidisers, 60, 92, 100
nitrogen
  cycle, 90
  nitrogen fixation, 51, 91
non-culturable cells, 40, 81
North Atlantic Deep Water, 7
numbers of bacteria, 32
nutrient cycles, 90
nutrient levels, water, 82

ocean basins, 2
oligotrophs, 38, 42, 134
outer membranes, bacterial, 82

panzootic, definition, 118
parasites, protozoan, 116, 121, 124
parasitism, definition, 9
pathogenesis, definition, 9
pathogenic mechanism, 118
pathogens,
  algae, 124
  bacterial, 115, 118, 123–4
  fungi, 116, 120, 124, 129
  rickettsia, 131
  viruses, 117, 122, 125, 130
pelagic habitat, 9
peptide antibiotics, 166–71
peritrophic membrane, 113
petroleum degradation, 149
Pfennig's medium, 26
pharmaceutical compounds, 157
phase contrast microscopy, 20, 79
phosphatic sediment, 8
phosphorus cycle, 99
photic zone, 6
phototrophs, 39
phycobilisomes, 51
phycoerythrin, 91
physico-chemical forces, 89

phytoplankton, 3
  definition, 8
picoplankton, definition, 9
picoplankton, 44
picophytoplankton, 39
plankton
  definition, 8
  net, 18–19
plasmids, 95, 119, 174
pleura, definition, 66
pleuston (see neuston)
polar front, 35
polymers, attachment, 174
polysaccharides
  antitumour, 171
  polyanionic, 164
  surfactants, 173
predator, 96
predator–prey relationships, 82–3, 88
pressure effects, 146
prey, 83
primary producer, 96, 148
primary productivity, 84, 99
productive zone, 6
prosthecae, 54
protozoan parasite, 116
psychrophile, 7
pycnocline, 6
pyrrole, 159–61

quinolinol, 165–6

radioactive tracer, 86
  transfer rate, 86
recycling, nutrients, 90
red-pest, 118
rickettsial disease, 131
roll tubes, 27
rotating drum sampler, 14

sampling methods, 12
  benthic communities, 13
  epibiotic communities, 13
  neuston, 13–14
  water, 13
sand, 8
scanning electron microscopy, 19, 89, 105, 113
seawater composition, 4–5
sediment composition, 8
selective isolation
  Aerococcus viridans, 129
septicaemia, 118
seston, definition, 9
settling, invertebrate larvae, 152
shake tubes, 27
shell-disease, 127
Shoyu, 153
siderophore, 119

silica gel medium, 137
siliceous sediment, 8
Simidu's medium, 23–5
single cell protein, 152, 174
sinking, 4
sporadic disease, 118
starvation–survival, 80
Sub-Antarctic Front, 35
submarine hydrothermal vents, 144
Subtropical Convergence Front, 35
sulphate reducers, 39, 59, 92–5, 97–8, 147
sulphur
  cycle, 93
  respiration, 93
surface zone, 6
surfactant, 173
survival, indigenous organisms, 79
  non-indigenous organisms, 82
'syringe array', 137

taxonomy, 45–78
  actinomycetes, 65
  aerobic Gram-negative bacteria, 55
  aerobic Gram-positive cocci, 63
  anaerobic Gram-negative bacteria, 59
  budding and appendaged bacteria, 54
  ciliates, 70, 74
  coryneforms, 65
  cyanobacteria, 51
  deep sea organisms, 141–6
  diatoms, 66
  dinoflagellates, 70
  facultative anaerobes, 58
  filamentous fungi, 68
  foraminiferida, 70, 74
  gliding bacteria, 53
  Gram-negative chemolithotrophs, 60
  Gram-positive endospore formers, 64
  methane bacteria, 62
  micro-algae, 66
  phototrophs, 47
  Prochloron, 52
  protozoa, 70
  radiolarida, 70
  spirochaetes, 66
  yeasts, 70
tetrabromopyrrole, 163
tetrazolium reduction, 80
thermocline, 7
thiosulphate citrate bile salt sucrose agar, 28
thylakoids, definition, 51
todorokite, 144
transmission electron microscopy, 41
trenches, 2
trichomes, 51
'tripods', 138
trophic level, 84
tuberculosis, 120

ulcer, 120
ultra-microbacteria, 80
upwelling, 4

vaccine, 119
valva, definition, 66
Van der Waals force, 89
vibriosis, 123
viral pathogens, 117, 122, 130
virulence plasmid, 119

warm vents, 135
Winogradsky column, 23

X-ray diffraction spectroscopy, 144

yeasts, 43–70

Zobell's 2216E medium, 23–4
zooplankton, definition, 9
zymogenous, 80